Aviation Computing
Systems

Other aviation titles published by McGraw-Hill

Piloting for Maximum Performance *by Lewis Bjork*

Optimizing Jet Transport Efficiency: Performance, Operations, & Economics *by Carlos E. Padilla*

Aircraft Safety: Accident Investigations, Analyses, and Applications *by Shari Stamford Krause, Ph.D.*

Airport Planning & Management, 3d Ed. *by Alexander T. Wells, Ed. D.*

Airport Operations, 2d Ed. *by Norman Ashford, H. P. Martin, and Clifton A. Moore*

Denver International Airport: Lessons Learned *by Paul S. Dempsey, Andrew R. Goetz, and Joseph S. Szyliowicz*

Flying Jets *by Linda O. Pendleton*

Kitplane Construction, 2d Ed. *by Ronald J. Wanttaja*

They Called It Pilot Error *by Robert Cohn*

Becoming a Better Pilot *by Paul A. Craig*

Redefining Airmanship *by Major Tony Kern*

Aviation Computing Systems

Mal Gormley

McGraw-Hill

New York   San Francisco   Washington, D.C.   Auckland   Bogotá
Caracas   Lisbon   London   Madrid   Mexico City   Milan
Montreal   New Delhi   San Juan   Singapore
Sydney   Tokyo   Toronto

Library of Congress Cataloging-in-Publication Data

Gormley, Mal.
    Aviation computing systems / by Mal Gormley.
        p.   cm.
    Includes bibliographical references and index.
    ISBN 0-07-032883-8 (pbk.)
    1. Aeronautics—Data processing.   2. Aeronautics—Computer network
resources.   I. Title.
TL563.G67   1997
629.135—dc20   96-41376
                                                                    CIP

McGraw-Hill

A Division of The McGraw-Hill Companies

1 2 3 4 5 6 7 8 9 0   DOC/DOC   9 0 1 0 9 8 7 6

ISBN 0-07-032883-8

*The sponsoring editor for this book was Shelley Chevalier, the editing
supervisor was Scott Amerman, and the production supervisor was
Suzanne Rapcavage. It was set in Century Schoolbook by
North Market Street Graphics.*

Printed and bound by R. R. Donnelley & Sons Company.

McGraw-Hill books are available at special quantity discounts to use as
premiums and sales promotions, or for use in corporate training programs. For more information, please write to the Director of Special
Sales, McGraw-Hill, 11 West 19th Street, New York, NY 10011. Or contact your local bookstore.

To my friend and wife, Pamela, for her love, kindness, emotional support, editorial suggestions, proofreading, endless patience, and humor, and because she believed in this project before I did; and to my two great kids, Katie and Tim, for their ideas, humor, and love.

Contents

Preface

"I find the greatest thing in this world is not so much where we stand, as in what direction we are moving."
OLIVER WENDELL HOLMES, 1858

As I was preparing this book, I was often asked what it was about. I thought the title was a dead giveaway, but I often got a quizzical look or a polite "Oh" from folks who asked. So here's the explanation: *Aviation Computing Systems* is a guidebook to the information technologies that have quietly reshaped the aviation industry in the last half-century and help make it possible for airlines to operate safely and efficiently, designers to develop and manufacture new aircraft, air traffic controllers to manage traffic, meteorologists to create accurate forecasts and distribute weather data quickly, military pilots to perform their duties effectively, and all pilots to plan their flights, maintain their proficiency, and keep up with what's going on in aviation.

We are rapidly approaching the time when data link and other advanced technologies will enable pilots, dispatchers, and air traffic managers to work together in real time to negotiate clearances, monitor weather, and communicate with each other anywhere in the world. In fact, some of these benefits have already started happening. For example:

- Data link technology is making it possible for Civil Air Patrol aircraft to relay, in near real time, aerial images of disaster areas so relief workers can respond appropriately.

- Airline flight crews crossing the Pacific Ocean are automatically reporting their position to controllers in Oakland, California, via satellite.

- Corporate jets crossing the Atlantic are capable of automatically sending waiting maintenance crews information about faulty hydraulic pumps.

- Fleets of advanced flight simulators are being used in real-time scenarios to develop the next generation of air traffic control computers.

- People everywhere are using the Internet to purchase airline tickets, find jobs in the aerospace industry, chat with other aviation professionals, preserve aviation history, and explore a host of other things.

Aviation computing systems play a key role in making these things happen.

If you're involved with aviation in any way—as a pilot, air traffic manager, airframe or avionics manufacturer, Federal Aviation Administration inspector, flight attendant, avionics or airframe technician, or even as a passenger—your future includes at least some of these aviation computing systems:

- Airline systems designed to provide passenger reservations, schedule flight crews, and maintain aircraft efficiently

- Flight simulators, flight training devices, and other computer-based training tools

- Military systems designed to effectively organize, coordinate, and conduct flight operations and maintain combat readiness

- Advanced communication, navigation, flight management, and cockpit display systems used on many modern civil and military aircraft

- Programs used by government agencies and private organizations to maintain and publish accurate pilot, aircraft, navigation, and safety information and to analyze new air traffic control systems

- Software that enables corporate flight departments to optimize their resources, project budgets, and analyze aircraft purchases

- Programs used by aviation enthusiasts to share news and views, "fly" advanced aircraft in simulation, and communicate with aviation luminaries

Aviation Computing Systems is a tour of the principal categories of aviation information technology and a glimpse of what each might have to offer in the near future. Also included are case studies, examples of some of the systems being discussed, interviews with industry insiders, and illustrations of aviation computing systems. *Aviation Computing Systems* also includes a directory of providers of aviation information technology and related resources.

Aviation Computing Systems was written for a wide audience, including

- Aviation industry professionals and enthusiasts who are seeking a greater understanding of an industry that's changing almost daily as a result of advances in aviation computing

- Pilots, flight attendants, technicians, airline reservations specialists and other personnel, air traffic controllers, and flight dispatchers who want to understand the role aviation computing plays in their day-to-day flight operations

- Computer and software developers who are considering the development of new products and services or opening a new direction in their careers

- Students considering a career in computing or aviation (the two are quickly becoming synonymous)

- Anyone intrigued with the industry and the promise of advanced computing applications

New concepts in aviation that are being developed all the time rely on advanced computing to make them possible. To highlight a few examples:

- Boeing's latest transport, the 777, was designed almost entirely on CATIA, a computer-aided design and manufacturing (CAD/CAM) program developed in France for a military jet fighter program. CATIA and other advanced systems like it have made it possible for aeronautical engineers to design an aircraft without using costly mock-ups and thousands of blueprints. At the touch of a light pen, engineers can test different design configurations to develop the most efficient way to route a mechanical, fly-by-wire, or fly-by-light flight control system through a wing assembly; improve the fuel efficiency of an engine, or add an easy-to-service galley to an older design.

- Regional carrier Comair has developed the Terminal Operations and Schedule Tracking (TOAST) system to integrate its flight planning systems, passenger flight information displays, and company communications networks at the carrier's new terminal in Cincinnati. The TOAST network was developed from off-the-shelf software for under $250,000 and in less than one year from requirements definition. That's remarkable in an age when a million-dollar budget for just about *any* business computing application is considered small!

- The cockpits of many modern aircraft are being fitted (and retrofitted) to take advantage of advanced microchip and computing technologies. New concepts and methods have enabled engineers to reduce avionics weight and cooling requirements while increasing their reliability, safety, and utility. By connecting the cockpit to advanced, high-speed airline networks, aircraft have become, in the words of an avionics engineer, "nodes on a mobile telecom network."

If your involvement with aviation involves only boarding an airliner for an occasional business or vacation flight, you'd probably be surprised at how much of the behind-the-scenes activity relies on microprocessors. Your airline reservation, for example, was probably booked by a travel agent using one of several computerized reservations networks. After booking your flight, your request for a special-diet lunch was routed to the airline's operations management computers and then to the catering company that serviced your flight.

The flight crew probably submitted their bids to fly that flight—a month or two prior to departure—by means of a computer system that sorts and tracks each crewmember's seniority, domicile, accumulated time, personal preferences, and recency of experience.

The airplane's maintenance status was evaluated by a computer system. Your flight's progress from gate to gate was tracked by a variety of flight-following and air traffic control (ATC) systems. Your aircraft's navigation system relied on the accuracy of a huge database of waypoints and provided the crew with a wealth of information unimaginable just 20 years ago. The flight crew was trained on simulators that provide startlingly realistic views and scenarios that could not be safely duplicated in flight.

And after you plucked your suitcase off the baggage carousel (another aviation computing-dependent system at many modern air terminals), it's likely that someone in the airline's front office examined your flight (plus hundreds of others) from a profit-and-loss perspective, using yet another air carrier-specific financial analysis program.

These examples are only a handful of the aviation computing applications. The world of aviation computing is evolving rapidly. It is my hope that *Aviation Computing Systems* will excite you about the subject, help you understand why it is developing so quickly, how you can benefit from and become involved in it, and, finally, what lies next.

This isn't a "gee-whiz" book. Instead, I hope you'll use it as a guidebook to an exciting, innovative industry-behind-an-industry that is reshaping aviation.

Introduction

*"Wilderness becomes Frontier becomes
Mainstream becomes Vision."*
                              ANONYMOUS

Your Tour Starts Here

There's been a revolution in aviation, and we've been too busy to notice it. Computers and software have turned aviation into one of the most dynamic areas in modern technology. So much in aviation today depends on accurate, high-speed, task-specific information processing that it's hard to imagine modern aviation without the use of computers and advanced integrated software. That's the subject of this book. *Aviation Computing Systems* takes you on a tour of the primary applications of electronic information processing in aviation, and, along the way, describes the benefits, opportunities, evolution of, and outlook for one of the most significant and largely unseen elements of the aviation industry.

The purpose of this book, at its most basic level, is to help you understand the role of computing systems in aviation. If you make your living—or plan to—in virtually any segment of aviation, you have been or soon will be affected by advances in aviation computing systems. It is essential that you have a clear understanding of the general design and roles that computers and information network systems play in aviation today. If you don't think so, you will soon be considered an anachronism by your peers, your fellow workers, and your employers.

I'm not just talking about whether or not you can turn on a desktop computer and use a word processor—the 1990s' version of a typewriter. Instead, I'm saying that if your work is in almost any aviation position, information technology, aviation law, or the investment community, you need to know about the kinds of specialized computer applications that are used to obtain, manipulate, and communicate information in aviation today and tomorrow.

The many applications of this knowledge to those who currently work or expect to work within the aviation industry will become evident as you read this book. If your profession is only on the periphery of aviation, I hope this book will make aviation a more understandable industry for you. It is an industry loaded with seemingly arcane terminology and more acronyms than most of us in the business can remember. This book can help translate many of these terms into understandable concepts, as well as introduce you to many concepts with which you might not be familiar.

Basic computer literacy is now often a requirement for almost all but the most menial tasks in aviation, simply because at some point in your workday you'll probably have to use a computer to gather or enter information or communicate with someone else. If you can demonstrate a familiarity with some of the technology used in the segment of aviation you plan to work in, you might surpass others competing for the same position. If you plan to work (or continue working) in any of the more specialized, technical areas of aviation—pilot, aircraft dispatcher, engineer, air traffic manager, weather specialist, avionics or powerplant maintenance, or even some specialized manufacturing positions—a working knowledge of one or more computer applications is mandatory in most cases.

Because so many elements of aviation seem to intertwine sooner or later, being aware of the computing systems used in areas adjacent to yours is beneficial. For example, if you are a corporate pilot, you are no doubt familiar with obtaining preflight information from a variety of computer sources. But what if you want to advance to chief pilot or flight department manager? Knowing the general capabilities of flight department management software systems, as well as online sources of information about the used turbine aircraft market, can put you ahead of someone who has been content with making flawless instrument landing system (ILS) approaches in the company Learjet 35. And using the Internet's aviation employment resources might make finding your next job a lot easier.

If you're planning a career in airline management, you'll naturally have to be familiar with airline reservation systems, operations control, and crew management packages. But where would you turn if you found yourself tasked with developing and implementing a company-wide crew resource management (CRM) program? Where would you turn if you were asked to be a participant in the development of a next-generation transport? (A growing number of manufacturers are using the Internet—and corporate intranets—to form teams of engineering and customer panels prior to and during the development of next-generation aircraft.)

If you're planning a career in aircraft engineering, knowing what software tools are being used to develop the latest commercial transports and military aircraft is going to be helpful, of course, but what if you find that you're more interested in creating airport terminals? What computer systems are being used to help keep passengers and crews informed? What do airport managers use to keep track of their tenants' leases? What if you had to create an airport where one didn't exist before—and it has to be operational in a week?

If you are in the investment world and need a quick, bird's-eye view of the aviation industry landscape, this book can definitely help. I have tried to provide you with a general knowledge of the major information tools and systems used in the aviation industry, as well as a basic understanding of many of the principal elements and much of the terminology of the aviation world in a real-world context.

If you are a writer in search of some background material for a novel or a nonfiction article on an aviation-related subject, this book could save you days of research and countless telephone calls because I've already done it for you.

If you are only marginally involved in aviation on a day-to-day basis—as a recreational pilot, aviation enthusiast, or just a frequent airline traveler—you'll benefit from reading this book. For example, recreational pilots can discover a variety of inexpensive and fun-to-use online services that can enrich one's weekend flying pursuits. Online discussion forums cater to the thousands of aficionados of desktop flight simulators. So what? Well, there's a growing possibility that using an inexpensive desktop simulator could reduce the actual flight time experience requirements—which equates to expense—for obtaining an instrument rating.

If you're only an infrequent flier, you might have marveled at how your ticket, your luggage, your special-diet dinner, and 250 tons of aluminum can arrive in the right place at the right time without bumping into another jumbo jet at 37,000 feet on a dark and stormy night. This book explains how much of that happens, day in and day out, with the help of advanced computers and the folks who use them.

Throughout the book I have tried to include descriptive examples and illustrations of the kinds of technology being discussed in each chapter. I have also taken nothing for granted about the reader's background and familiarity with aviation terminology, so I kept the discussion on as basic a level as possible so that individuals with no aviation training or qualifications can follow the discussion easily.

Let's face it. This isn't a European vacation travelogue or a Tom Clancy novel—it's a tour of some highly complex, technical concepts and subjects, but I've tried to keep it as readable as possible. Wherever possible, I let the people I interviewed describe the subjects they know best so you can get a sense of their interest and enthusiasm.

As I researched this book, I discovered that no single book could properly describe every last detail of an industry that is constantly reinventing itself. However, I hope this book alerts you to the primary elements that, for now, appear to be driving the future of aviation.

Chapter-by-Chapter Summary

The following descriptions provide a thumbnail sketch of the range of subjects covered in each chapter. I hope you notice that there are a number of recurring themes and subjects. Among the most significant of these is data link and the notion of the airborne data terminal. It is remarkable how many people I spoke with in researching this book mentioned one or both of these subjects, sometimes in the same breath. It would seem that the cockpit is becoming a workstation with wings!

Chapter 1: CAD/CAM: The virtual prototype revolution

Sophisticated software has revolutionized the entire engineering industry, perhaps most spectacularly in the aircraft manufacturing segment. Chapter 1 describes how the engineer's drawing table and slide rule have been replaced with an arsenal of sophisticated electronic aids. This chapter describes the range of computer-aided design, engineering, manufacturing (CAD/CAM), and related engineering tools that help airframe manufacturers design and manage products from initial conceptual sketches to finished product, including subassembly specification and integration, parts ordering, product data and document management, shop-floor management, and product marketing. These programs create virtual prototypes that have completely eliminated the need to produce costly three-dimensional models of the finished product, thus saving manufacturers a fortune in complicated parts-matching and redesign costs. This chapter features case histories of CAD/CAM applications to some of today's most advanced aircraft and interviews with leading airframe engineers.

Chapter 2: Flight-deck computing systems

Almost overnight, the flight decks of modern aircraft have become airborne computer centers. In a span of less than 20 years, the electromechanical and vacuum-powered displays and instruments of earlier airline, military, business, and general aviation aircraft have been replaced with "suites" of integrated digital electronic sensors and displays that resemble, in form and function, advanced computer systems. In Chapter 2 I explore the development of "glass cockpit" avionics, in which software development now constitutes between 75 and 90 percent of the cost of the systems. The discussion includes the development and implications of data bussing in civil and military avionics; visual display systems; the marriage of cockpit avionics to aviation information networks; the development of the Ada programming language; development challenges facing avionics developers and certification agencies; engineers' new definitions of the functional packaging of their systems; the impact of technologies such as satcom; airline cabin inflight entertainment and information systems; data link; and the outlook for further development of what some are calling "the airborne data terminal."

Chapter 3: Wired wings: Networks, nav databases, and the Internet

The way the media reports it, you'd think the Internet was invented yesterday, and nothing preceded it. Aviation has been a major user of global networked services back to the very beginnings of aviation. Indeed, the Wright brothers used a network "site" at Kitty Hawk to notify the world of their success in December 1903. And so it has been ever since. In this chapter we'll take a look at some of the primary nets used by the commercial operators of the world since the 1930s and 1940s—ARINC and SITA, respectively—and the potential these have for making aircraft even more productive for their operators and passengers.

Chapter 3 also explores how aviation found the Internet and is making it one of the most powerful resources in aviation history. This chapter describes some of the many uses of the Net for the airlines, military, organizations, employment services, original equipment manufacturers (OEMs), educational institutions, classified advertisers, publishers, and individuals. Along the way, I briefly explain what the Internet is, some of its terminology, and how you can use this resource for business, education, and recreation.

Talk about navigating the Net! I also take a look at a major powerhouse in the aviation industry: Jeppesen, whose navigation databases are among the most crucial—but largely unseen—advances in aviation safety. Jeppesen's electronic databases and those produced by various civil and military agencies are largely what make it possible for aircraft to navigate precisely. Other databases used in aviation are also discussed in this section.

Chapter 4: Virtual flight: Simulators and computer-based training

With the exception of avionics, I doubt if there is any other area in aviation in which computers and flight are so closely intertwined as the modern flight simulator. Providing highly realistic sights, sounds, cockpit displays, and physical motion requires some of the most advanced computing technology there is. Chapter 4 explains how FlightSafety International, a leading flight simulator maker and training provider, produces the level of fidelity required to meet the FAA's Level D standard. In this chapter I take a look at the differences between flight simulators, flight training devices, and other forms of computer-based training, as well as some of the important issues this capability is raising in the aviation community.

No tour of the simulation industry would be complete without mentioning the NASA Langley Research Center's Simulation Systems Branch Facility in Hampton, Virginia. The researchers at this facility operate eight flight simulators that are used for a variety of fascinating research programs.

Pilots of most cabin-class business aircraft, as well as a growing number of general aviation (GA) pilots regularly attend refresher training programs that include use of highly realistic (computer-guided) flight simulators. Recently the use of computer-based multimedia systems in the classroom is augmenting, if not replacing, the use of overhead and slide projectors. And what business jet captain, weekend GA pilot or pilot-wannabe hasn't daydreamed of piloting an F-16 Eagle, a Spitfire, or a space shuttle? The immense popularity of PC-based flight simulators has generated magazines, online service discussion groups, Internet Web sites and conventions. I take a closer look at simulation, computer-based training, and simulation gameware in Chapter 4.

Chapter 5: Air carrier computing

Computers have enabled airlines to automate many tasks, and new applications are being developed all the time. This chapter looks at the primary functions of computers in the airline industry—reservations systems, flight planning and dispatch, crew scheduling, maintenance management, and cargo management, as well as how flight information display systems (FIDS) are becoming a model for airport intranets. Included is a short history of the most respected and emulated of the reservations systems—SABRE, which was recently spun off by its parent company, AMR Corporation.

Chapter 5 also examines how the airline's use of data link technology is creating new forms of information media and new opportunities for airframe and avionics manufacturers and others that are beginning to offer life-cycle maintenance support to their customers.

Despite the airlines' use of computer technology, one gets a mixed impression about the industry's use. Some carriers, such as American,

United, Delta, British Airways, Singapore, and a few others, appear to be on the cutting edge of computer technology and are willing to invest heavily in advanced systems in search of additional profit or market share. At the same time, other carriers turn their computing tasks over to outside computing and network services either because they lack the expertise in-house or they have chosen to concentrate on their primary stock-in-trade: moving people and cargo by air. Industry insiders claim that some major carriers might be too hamstrung by their massive information technology departments to respond rapidly to market changes and the needs of their end users.

Chapter 6: Military aviation computing: The challenge of information warfare

Computers have played a role in military aviation almost from the outset of the computer age during World War II. Although much of what is discussed elsewhere in this book equally applies to military aviation computing, it is worthwhile to take a brief look at how this segment of aviation has harnessed the power of the computer. One issue of growing importance is the subject of information warfare. To paraphrase a popular maxim, "Whoever has the most information wins."

Like many organizations, a cash-strapped Pentagon is looking for commercially available off-the-shelf solutions (COTS) to improve its informational advantage over its adversaries. Along with greater reliance on stealth technology and small, stealthy, unmanned aerial vehicles (UAVs) for reconnaissance (which owe so much to advanced computing technology), the military is always looking to improve or replace its legacy data collection and communication resources, as well as study how secure, high-bandwidth, private communications networks and satellite links will be the "weapons" in this new battlefield.

In Chapter 6 you'll also meet an airborne information specialist who describes how computers are used in front-line, airborne applications, even if the front line is a simulated air battle.

Chapter 7: Business and general aviation computing

Depending on your aviation background, business aviation and general aviation can be viewed as either one or two distinct entities. The same can be said about the aviation computing products and services used by each of these segments. Business and general aviation computing products span a wide range of functions and capabilities that have been slowly converging as each segment evolves. Chapter 7 discusses the types of systems that each segment uses and provides some illustrative examples.

This chapter includes an interview with Dick Aarons, one of the first people to recognize the value of personal computer use in business aviation. He formed one of the first companies to market business aviation flight department software. This field is still surprisingly fragmented, with many vendors that have been providing their products and services for years. But the business and general aviation computing software marketplace is equally open to the entrepreneur who can find a niche and fill it with a worthwhile product or improve an existing one.

Chapter 8: Air traffic management systems: At a turning point?

Chapter 8 takes a look at the FAA's development of air traffic management systems from a simple radio and teletype circuit to today's labyrinth of networks and airspace fiefdoms. Today, the agency's computer systems that are used to monitor and manage air traffic are marvels of computer engineering technology. Yet even the FAA admits much of its hardware and software is woefully out of date or has been patched together for years, and it is time for a major overhaul. Some of the more promising information technology solutions are described here. The FAA also recognizes the need to accommodate a rising number of airspace users and to do so in a way that removes the antiquated operational constraints of "positive control" in light of today's available technology and aircraft capabilities. The aviation industry has coined the term *Free Flight* to describe an emerging concept for the nation's airspace that offers the prospect of greatly increased operational flexibility for airspace users. But Free Flight, as proposed by the FAA and others, is not without its critics, who argue the goals of Free Flight could be achieved more quickly and inexpensively with only a slight shift in strategy. Only time will tell what the next ATC configuration will be, but it is nearly certain it won't resemble its current form. This chapter provides you with a bird's-eye view of the systems, history, issues, and possible answers to today's airspace challenges. And, in case you're wondering, data link plays a major part.

Chapter 9: The weather net and Uncle Sam online

How do you keep track of 150,000 pilots and thousands of aircraft? How do you make sure civil and military flight crews have the latest weather forecasts? How should you communicate with an agency made up of tens of thousands of employees in 12 regional offices and a couple dozen other facilities? This chapter explains how! Although ATC gets a lion's share of the FAA's computing budget, other parts of the FAA and other federal agencies with aviation-related information technology systems are also important elements of the aviation computing picture. One of the largest segments is the government's weather service providers. In this chapter you also learn how weather information is gathered and exchanged, both in the United States and elsewhere. Along the way, some of the more promising areas of information systems development in weather forecasting and dissemination are described.

Chapter 10: Data link: Tomorrow's digital airways

The next step toward creating a seamless, digital aviation environment is the implementation of data link communications technology. Data link will finally connect the ground side with the aircraft in real time. Data link, say its proponents, is going to become the cornerstone of much of aviation's future. By introducing a new way to communicate, data link will provide significant benefits to all airspace system users. Proponents of data link technology—primarily the FAA, the International Civil Aviation Organization (ICAO), and the OEMs, say that data link implementation, over time, coupled with advances in functionality, will lead to increased safety, efficiency, and capacity. These improvements in airspace management will contribute to increased savings in aircraft operational costs, savings that are estimated in the millions of dollars. Data link is also considered one of the fundamental technologies necessary to achieve Free Flight. Data link applications will also include (1) ATC services; (2) flight information services; (3) surveillance services; and (4) advanced operational applications, such as inflight maintenance fault warnings.

One of the most significant communication, navigation, and surveillance/air traffic management (CNS/ATM) initiatives in a generation is FANS (Future Air Navigation System). This international effort, which is described in Chapter 10, is likely to have far-reaching implications for aviation computing systems, especially avionics, air traffic management, and airline operations. It relies heavily on data link and satellite-based navigation technology—and a lot of international cooperation—to make it happen.

Hopefully, by the time you've read this far, you have a new appreciation for the scope of information technology in aviation today, and a better understanding of where it might be headed. It is my hope that this information will be of value to you in your aviation career or as someone with a desire to understand how it all works.

Appendix A: Aviation computing system resources

Appendix A lists some 100 sources of aviation computing-related products and services made by companies whose primary business is within the aviation community.

Appendix B: Aerospace Internet hot sites

Appendix B shows you how to find what's hot on the Internet. It lists URLs (universal resource locators), which are those electronic addresses everyone's using. I've visited many of these places, and I provide a brief description of each. By the time the next edition is published, you'll probably have your own Web site. If it's cool enough, I'll add yours to the list.

Glossary

Any book on something this technical has to have a glossary to help you keep track of all the new terms and acronyms.

On the History of Aviation Computing . . .

While I was researching this book, I quickly realized that the history of aviation computing isn't well documented. This might be because the subject is still so new and things change so quickly. Nevertheless, as you proceed through the book, you'll find a number of historical perspectives, such as the development of ARINC, Jeppesen-Sanderson, early air traffic control, the Ada computing language, and a few other subjects. The development of aviation computing closely parallels the development of computing itself, and I'll be bold enough to state that a number of major advances in the computer industry have come about largely because of aviation's need for high-speed, real-time computing. Examples of this include military ballistics (which led to EINIAC, the world's first mainframe computer), airline reservations (SABRE, et al.) and flight planning systems, network communications for military intelligence, aviation weather and air traffic management purposes, flight simulation, and satellite communications. If you have material on the history of aviation computing (hardware, software, documentation, anecdotal information, etc.), contact me or donate it to an aviation historical archive (such as the Smithsonian) so that it can be properly collected and preserved. Also errors and omissions occasionally occur during the editing process. Please feel free to contact me via e-mail at: comet@lincoln.midcoast.com.

Mal Gormley

Acknowledgments

My sincere thanks to all those who helped me write this book. I am especially grateful for the encouragement of Richard Aarons, editor-in-chief of *Business & Commercial Aviation*. Dick's support helped me become a specialist in the topic of aviation computing technology. I also owe a debt of gratitude to Gordon Gilbert, *B/CA*'s senior news editor, for his wit, wisdom, and many helpful comments and suggestions. Also, fellow writers Bob Searles and Bob Parke, for their encouragement and advice. And thanks also to the many others who were instrumental in the development of this book, including Richard Reinhart, Walt Shiel, Shelley Chevalier, Neil Krey, Preston Henne, Peter Wanner, Ronald Houde, Russ Williams, Bill English, Jim Veideffer, and Jessica Salerno.

CAD/CAM: The Virtual Prototype Revolution

When one thinks of aircraft engineering, the image of hundreds of engineers sitting at row upon row of drafting tables with slide rules in hand might come to mind. That picture is totally out of date, I discovered while researching this corner of the aviation computing world. In fact, it is no longer news that most aircraft engineers have replaced their drafting boards and slide rules with an arsenal of powerful computing tools. Sophisticated software has revolutionized the entire engineering industry, but perhaps most spectacularly in the aircraft manufacturing segment. Many of the leading developers of engineering software proudly point to their aircraft design applications as distinguished examples of their capabilities. After all, how much more demanding an environment can there be for an engineering system than the development of a commercial airliner or a stealth fighter?

Today a whole range of computer-aided design, engineering, and manufacturing (CAD/CAM) and related engineering tools can help design and manage products from initial, conceptual sketches to finished product, including subassembly specification and integration, parts ordering, product data and document management, shop-floor management, and product marketing. These programs create virtual prototypes that completely eliminate the need to produce costly 3-D models of the finished product, thus saving manufacturers a fortune in complicated parts-matching and redesign costs. Computational fluid dynamics (CFD) aerodynamic modeling systems allow engineers to squeeze every last iota of aerodynamic efficiency from their designs by visually displaying the interaction of airflow with the aircraft's surfaces. Computational stress design (CSD) programs make it possible to determine the optimal structure for each and every part used on an aircraft to maximize its strength while minimizing its weight. These tools allow designers to extend the range and altitude capabilities of existing designs or explore new and unconventional design concepts.

But as I also discovered, engineering software has other applications within aviation. For example, CAD/CAM systems are being used by civil and military aviation authorities for a variety of aviation-related applications, including

- GIS (geographic information system) applications to create detailed airport and airspace maps, aeronautical charts, and air traffic control management planning aids

- The development of 3-D instructional tools used at the FAA's training academy in Oklahoma City, where the nation's air traffic controllers and administrators receive initial and recurrency training

- Tracking and management of 11 million square feet of space and leases at Boston's Logan International and nearby Hanscom Field

- The design of "instant" airports in combat zones

It's no wonder then, that CAD/CAM has become a multibillion-dollar business. In the first part of this chapter we'll take a look at some sterling examples of how the industry is using these tools to develop aircraft faster and more economically. Case histories and thumbnail descriptions include commercial airliners, high-tech, "stealthy" military designs, globe-girdling corporate jets, and even homebuilt recreational aircraft. Part of our "tour" includes some sound career advice and guidance for engineering students from some of today's leading aerospace engineers. Then we'll take a look at some examples of other CAD/CAM applications to illustrate how computers are revolutionizing the aviation industry outside the periphery of traditional aircraft design.

Computerized Design, Engineering, and Manufacturing

Big-name aircraft OEMs (original equipment manufacturers) such as Boeing and Lockheed Martin have led the way in the application of computer design and manufacturing tools and have made them the standards of the industry. But other manufacturers, less well known to the public, also have been taking advantage of the power, time, and cost savings these tools provide.

The introduction of software tools that enable engineers to design an entire aircraft without the use of costly mock-ups has been transforming the aviation manufacturing industry for the past 20 years or so. The first engineering programs were developed for use on large mainframe computers in the 1960s, later migrating to minicomputers in the mid-1970s. Only very recently have programs robust enough for the demands of complex aircraft engineering tasks begun to appear on networks of personal computers.

Programs in this category are now enabling engineering teams thousands of miles apart to develop ideas quickly and see how their concepts will appear in finished form. Designers and engineers can quickly develop internal and external structures, mate airframes with powerplants, route control cables, thread hydraulic lines and avionics wiring bundles, and place sensors and display systems, appliances (fans, lights, toilets, galley equipment, seats, etc.) in different locations for best fit. Structures can be quickly developed and tested in different ways to produce the ideal balance between strength and weight—a constant battle for engineers. Computational fluid dynamics software tools make it possible for engineers to visualize airflow patterns across the wings, empennage, fuselage, engines, engine mounts, and other external surfaces. This ability allows engineers to experiment with different airfoil shapes without requiring models or costly wind-tunnel testing. Hundreds of different shapes can be tested for best fit and airflow paths that, for example, maximize an aircraft's range or enhance its high- or low-speed flying characteristics or even its noise signature. By means of computer animation, the latest versions of CAD/CAM systems are enabling engineers to visualize how to improve access to the aircraft for routine maintenance tasks. Still other programs are making it possible to produce and distribute the intricate instructions for assembling the aircraft on the factory floor or reduce scrap produced in fabricating the costly materials used on today's transports.

The Boeing 777

One of the first companies to capitalize on the CAD/CAM revolution is Boeing Airplane Company, based in Renton, Washington. In the 1980s, Boeing began developing a computer design, manufacturing, and support strategy that would take the company into the twenty-first century. After trying a number of pilot programs, the company decided to invest in CATIA. Developed by Dassault Systemes in France and marketed by IBM in the United States, CATIA, along with several Boeing-developed engineering applications, enabled Boeing engineers to simulate the geometry of an airplane design on a computer—and eliminate the need to construct costly mock-ups. CATIA and other similar programs provide an electronic definition of a part or collection of parts that can be manipulated in 3-D on a computer monitor. Computational fluid dynamics (CFD) enables engineers to take that part's geometry to create a surface on which to perform an analysis of how fluid (air or water) will flow over that surface or to analyze the structure.

One of greatest obstacles in designing and engineering a complex machine, whether it is an airplane, a ship, or even miniature surgical equipment, is the difficulty in properly fitting the numerous parts together during assembly. By 1989, Boeing engineers were confident that they could significantly reduce the costly rework caused by part interferences and fit problems by digitally preassembling the airplane on a computer. The improved accuracy in part design and assembly, as well as the instantaneous communication made possible by this technology, convinced Boeing that the steep initial investment required to make it possible would pay off in better airplanes and reduce new-product development-to-rollout cycle times.

The opportunity to apply the new CAD/CAM approach, as well as other new engineering and manufacturing ideas, came in October 1989 with the launch of the Boeing 777 twinjet wide-body transport project. The 777's two variants, the –200 and the –300 (the latter launched in June 1995), seat from 305 to 440 passengers and 368 to 550 passengers, respectively, and are designed to fill the niche between the com-

Figure 1.1 After trying a number programs, Boeing decided to invest in CATIA software for its 777 program. Developed by Dassault Systemes in France and marketed by IBM in the United States, CATIA, along with several Boeing-developed engineering applications, enabled Boeing engineers to simulate the geometry of an airplane design on a computer—and eliminate the need to construct costly mock-ups. The 777 first flew on June 12, 1994. (Boeing)

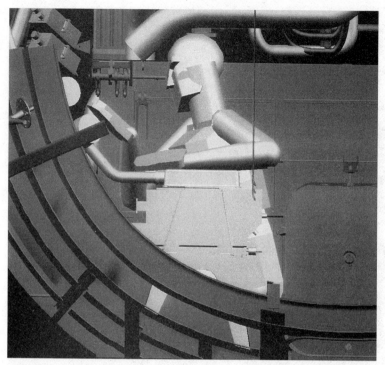

Figure 1.2 "CATIA Man" goes to work. Boeing engineers used 3-D "human" models to ensure airline mechanics have access to functional working areas of the 777. (Boeing)

pany's 767–300 and 747–400 models and to replace earlier models of the 747. The 777–200's first flight came on June 12, 1994, and the first revenue flight took place on June 7, 1995. The 777–300 is expected to roll out in August 1997 and enter service in May 1998. The –200 is available in a standard version capable of flying, depending on takeoff weight, configuration, and engines installed, from 3780 nautical miles (nm) to 4630 nm or in a "stretch" version with a range of anywhere from 5960 nm to 7230 nm, depending on configuration and power-plants installed. The two models have between 132,500 and 150,000 unique parts. Add in the rivets, bolts, and other fasteners required to hold one of these immense aircraft together, and the count grows to more than three million parts.

Following the definitions of the basic configuration (which was defined mostly by air carrier operators around the world), the launch of the 777 program established design/build teams to develop each element of the airplane's numerous structures and systems. This new approach enabled all of the specialties involved in the airplane's devel-

Figure 1.3 Boeing design/build teams members used sophisticated workstations to design and electronically preassemble the entire Boeing 777. The team currently uses about 1700 workstations connected to the world's largest mainframe computer cluster. (Boeing)

opment—some 238 teams of designers, manufacturing representatives, tooling experts, engineers, financial analysts, suppliers, certification authorities, customers, and others—to work together to create the airplane. Based at the same location, team members worked concurrently, sharing their knowledge, instead of applying their skills sequentially.

To help conduct this symphony of talent, four IBM mainframe computers were networked into a cluster in Boeing's facilities in Washington. This cluster was in turn linked to mainframes in Wichita, Philadelphia, Japan, and elsewhere.

But it was CATIA that drove the project. From the beginning of the 777 program, the three major participants in the system—Boeing, Dassault Systemes, and IBM—developed and signed a simple but profound agreement, stating the commitment of the three companies to work together to deliver 777 products and services on schedule.

But CATIA was not enough. The possibilities such a system offered were stunning but required a total rethinking of the processes involved in building airplanes to gain the most leverage from using the new technology. A vital part of the effort to ensure a service-ready 777 at delivery was the development work conducted in Boeing's Flight Controls Test Rig and Systems Integration Lab. These facilities, and others located in the Integrated Aircraft Systems Lab, permitted integrated testing of the 777's systems in simulated flight as well, before the aircraft flew. On previous aircraft development programs, the avionics, mechanical systems, engine controls, flight control system, electrical system, hydraulics, and other systems were given their first fully integrated workout on the day the aircraft first flew. By testing the "virtual" airplane, this milestone was reached earlier in the development process, allowing for more effective flight testing and a smoother service introduction.

Boeing applications enhanced CATIA systems in three major areas: data management, user productivity, and visualization. Each of these major enhancements was required to deal with the size and scale of productively managing the millions of aircraft parts that go into a 777 modeled on CATIA.

Once all the computing applications were in place, Boeing engineers and designers were able to use the 3-D digital tools to see parts as solid images and then simulate the assembly of those parts on the screen, easily correcting misalignments and other fit or interference problems that cropped up along the way.

The effort paid off. As a result of the investment in computer solutions, the 777 program exceeded its goal of reducing changes, error, and rework by 50 percent. Parts and systems fit together better than anticipated and at a higher level of quality than was previously possible. To give an example of this accomplishment, Boeing engineers like to point out that while most airplanes line up to within ½ inch, the first 777 was just .023 inch—about the thickness of a playing card—away from perfect alignment!

In June 1995, the Boeing 777 Division was recognized for its innovative application of computing technology to the 777 program when it won the top spot in the Manufacturing category of the annual Computerworld Smithsonian Awards. The awards honor the world's most creative and innovative uses of information technology that benefit society. By earning top honors in its category, the Boeing Computing and Design Application earned a place in history in the Smithsonian's permanent research collection. Information on the application also is available at *The Information Age: People, Information and Technology,* a permanent exhibit at the National Museum of American History in Washington, DC.[1]

Lockheed Martin

When you say "Skunk Works" to aircraft engineers, they immediately think of the super-secret facility in the western Mojave desert, founded

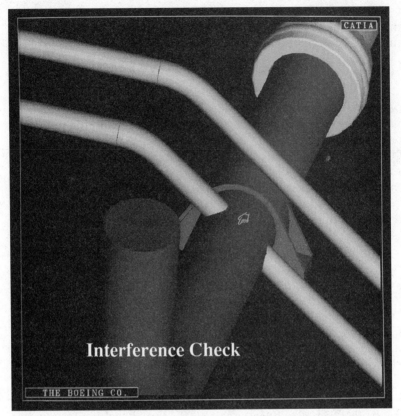

Figure 1.4 CATIA and other CAD/CAM programs allow engineers to find and correct part interferences, such as this one discovered during Boeing 777 development. (Boeing)

in the days following WWII by Kelly Johnson, one of Lockheed's top engineers. The Lockheed Martin Skunk Works continues to create some of the world's most sophisticated military aircraft, including the U-2 high-altitude spy plane; the world speed-record-holding SR-71 Blackbird reconnaissance jet; and the F-117 NightHawk stealth fighter. The company is responsible for a number of other "dark" projects that haven't been declassified yet.

Recently, the company used CATIA and IBM RISC System/6000 workstations to design and build "DarkStar," one of the world's most advanced unpiloted reconnaissance aircraft, in a fraction of the time and resources normally required for a project of this magnitude. DarkStar was designed for fully automatic, all-weather takeoff, flight, and recovery and will operate at altitudes in excess of 45,000 feet for eight-hour periods or longer. Its 8600-pound gross weight is packed into a sleek, low-profile design. Its height is just 5 feet, and its length is 15 feet, but it has a wingspan of some 69 feet, making the aircraft resemble a boomerang. DarkStar's sensitive radar and electro-optic sensors are designed to provide the military with wide-area surveillance for bomb-damage assessment and detection of enemy missile systems in near-real time.

The first of two DarkStar technical demonstrators was rolled out in June 1995, just 11 months after a team of Lockheed and Boeing designers started the project. (The aircraft was slated for its first flight in late 1995.) Similar projects have taken years and hundreds of designers to complete.

"Our team approach involves networks of colleagues and suppliers working off the same aircraft model in a virtual enterprise," said Josh Levinski, manager of programs and technology for Lockheed Martin Skunk Works. "The IBM RS/6000 allowed wing designers at Boeing in Seattle and fuselage designers at the Skunk Works in Palmdale, California, to simultaneously access and collaborate on specific designs. The result speaks for itself: a perfect fit in the first assembly." Precision throughout the building process was especially critical because DarkStar is made mostly from composite materials instead of metal. Tooling began from the CAD models in just two weeks. As testimony to the precision that computer-aided engineering makes possible, the process of joining the wings, which were made in Seattle, to the fuselage, which was made in Palmdale, took only a few hours, and the design required no reworking, which surprised even the engineers.

DarkStar is one of the first military aircraft to have been designed without the creation of physical mock-ups, which reduces cost and speeds the design process. For example, radar cross-section analysis on the CATIA model gave designers assurances that the aircraft would have an appropriately low, or "stealthy," radar profile—a major requirement for a modern reconnaissance aircraft. And maintainability was assured by CATIA programs, which determined whether DarkStar technicians can access particular areas in the aircraft—with or without a glove.

The IBM RS/6000 server also supported a network of PCs on the factory floor, running resource planning and quality assurance programs. CAD models, for example, could be sent directly to the factory floor, giving builders real-time access to data required to perform manufacturing processes.

Bob Fischer, chief engineer for the DarkStar program, underscored Levinski's comments, saying that the next project would only call for minor improvements in CATIA's capabilities. "We'd like the design to be more 'paper-less,' " said Fischer. "Outside of that, I'd go with the same program."

Figure 1.5 Lockheed's stealthy DarkStar unmanned aerial vehicle (UAV) was designed with CATIA software. This step enabled engineers to produce the first aircraft in less than a year. Although this aircraft crashed on its second flight, engineers were confident their design is basically sound. (IBM)

Fischer explained how such a complex undertaking is managed. "You put your design under a configuration control log, which [starts at] the time you release your design to Manufacturing, QA, and Procurement, and it states the configuration revision of every drawing that you have. And we and other companies use a 'redmark' system, which has a tracking logo that is recorded. We keep track of that as we update drawings.

"Another good thing about using an electronic database—the number of changes has been greatly reduced, because [the design] is verified before it's released," said Fischer. "The interference problems are almost nonexistent. The thing you still have to worry about is system functionality, where you have to integrate the hardware and the software [in the design]. That's where you still have to work your way through it."

Levinski would like to see improvements in interference-elimination processes, especially between subassemblies and on the aircraft as a whole. He's also eager to see improvements in developing a common parts list that meets the needs of both engineering and manufacturing departments. "We're trying to achieve an 'as-designed-as-built-as-delivered' parts list. That's been a big bugaboo of the aviation industry," said Levinski.

CATIA and other programs haven't been able to completely eliminate the need for wind-tunnel testing of new shapes, since there isn't a program that can accurately simulate the way wind vortices swirl about an aircraft—yet. "We're just beginning to get that door to swing open, but we're not there yet," said Fischer.

Both Levinski and Fischer concede that there might be faster, more nimble programs on the market than CATIA, and one might eventually replace it, but none has proven itself equally capable of managing million-part inventories in an enterprise-wide setting the way CATIA has. Further, to import a new program into an organization as large as Lockheed could create more problems than it solves, when issues of training, industry-unique applications, and enterprise- and third-party product integration are considered.

There's been occasional criticism of CAD/CAM programs in that they require a lot of time to master or that managers of CAD/CAM teams don't always understand what the designers are talking about because CAD/CAM is an arcane science. Fischer, however, thinks such criticism isn't valid. He says that training a new hire on a CAD/CAM system doesn't require any more training that mechanical drafting tools used to require. Indeed, on the DarkStar program, 30 of the 36 designers had no prior experience in developing a prototype vehicle, and many of them had to receive training in a very short time to learn how to manipulate the CATIA design tools. Their first drawing might have taken an extra day or two, said Fischer, "but by the time they turned out their sixth or tenth drawing, it was as quick as an engineer with ten years' experience with it. Based on my experience, there is no slowdown just because the engineer has never seen a computer before."

Fischer is impressed with the level of computer skills of today's students in general, but he believes engineering schools could do a better job of teaching fundamental kinematics, mechanical drafting, and descriptive geometry. "You still should know in your mind how to complete a view." Some schools, such as UCLA and the University of Kansas, have recognized this need and are beginning to offer this training again, said Fischer. He also would like to see more schools teach students to use some of the basic computer engineering tools and databases used in aviation manufacturing. Fischer lauds schools that encourage students to explore the practical applications of engineering by creating working prototypes of their designs.

According to Fischer and others, today's new design tools provide a powerful resource for engineers who might, for example, have difficulty turning or rotating new designs in their heads. And not having to produce mock-ups has shaved six months off development time. "Computerized CAD/CAM," said Fischer, "is a tool that has gone not as far as it needs to go—basically it's in its infancy—but people have no qualms about using it."[2]

G-V: Gulfstream's business jet

During a recent visit to Gulfstream Aerospace Corporation's main facility in Savannah, Georgia, as part of an assignment for *Aviation Week & Space Technology*, I had an opportunity to see firsthand how computers are changing the business of designing high-performance aircraft. The company was about to fly its latest model, the G-V, for the first time. For the previous three years, Gulfstream engineers had been building and checking its parts using 3-D electronic mock-ups.

Gulfstream has been producing business aircraft for 37 years, beginning with the Gulfstream G-I, a pressurized, twin turboprop that first flew in 1959. In 1966, the company introduced its first business jet, the G-II. The G-II became the first corporate jet capable of carrying 14 to 16 passengers, plus crew, for long distances (maximum range was 2613 nm) at "airline" altitudes, and several highly modified versions are still used by NASA to train space-shuttle astronauts. The G-III was certified in 1980. It featured increased thrust and range (3691 nm maximum). The G-IV derivative received its FAA certification in 1986 and became one of the first truly intercontinental business jets, with a maximum range of 4220 nm at Mach 0.78, with three crewmembers, eight passengers, and industry-recognized fuel reserves. Various derivative models appeared along the way, including the G-IIB and the G-IVSP, which featured various improvements. The company's latest model, the G-V, took its first flight in November 1995.

The G-V was developed to be the first "ultra long-range" business jet, with design goals that include a range of 6500 nm with eight passengers, a crew of four, ample fuel reserves, a certificated cruise maximum cruising altitude of 51,000 feet, and enough takeoff performance to allow the airplane to depart high-altitude airports in hot weather without worry about runway length requirements. The range capability would enable passengers to depart from New York and fly nonstop to Tokyo, a feat never before accomplished in a business aircraft. The passengers aboard such an airplane will be the senior executives of some of the world's largest corporations, government VIPs, and top-name entertainers who regularly need to travel such distances. Achieving these design goals would put the airplane well ahead of its nearest competitor, the Bombardier Global Express, an airplane with nearly identical design goals. At the time of my visit, the Global Express (often referred to as the GX) was at least a year away from its first flight, but the competition between Gulfstream and Bombardier was, and continues to be, fierce.

Figure 1.6   The Gulfstream G-V, also designed with CATIA, is being developed to be the first "ultra long-range" business jet, with design goals that include a range of 6500 nm and a certificated cruise maximum cruising altitude of 51,000 feet. (Gulfstream)

Over the course of two days, Gulfstream's engineering and G-V flight-test teams and senior executives proved to me that their new airplane would meet their stated design goals. A tour of the facility also gave me a close-up look at the use of typical aviation computer design. At many of the engineering workstations I saw, computer monitors depicted elements of the aircraft in phosphorescent green, purple, yellow, and white. The company had invested some $20 million in CATIA solid-modeling software and related hardware to carry it into the twenty-first century. The investment appears to be paying off.

CATIA and a number of Gulfstream-developed programs enabled engineers and designers to work concurrently on all of the problems and complications of building the G-V's parts without the time, cost, and effort of actually constructing prototypes. The computerized mock-ups enabled Gulfstream engineers to check the size, shape, and fit of parts as if they had already been built, eliminating the traditional separation between the design and manufacturing processes. As a result, Gulfstream was able to simulate many aspects of the airplane before the first piece was assembled, making it possible, for example, to assemble the first unit's fuselage 34 days earlier than planned.

The use of CATIA began when Gulfstream executives met with representatives from IBM and Dassault Systemes. The initial CATIA work was done on Gulfstream's mainframe computer but soon spread to more than two dozen IBM RISC System/6000 workstations. CATIA converted each part of the airplane into 3-D mock-up form, from wing contours to circuit breakers to landing gear doors to galley.

I saw other applications of computer engineering and design tools in Gulfstream's Structural Test and Integration Test facilities. The engineers in the Structural Test department, for example, were busy verifying the G-V's structural integrity under simulated maximum anticipated stress conditions. A recent upgrade of the system had tripled the number of tests that could be accomplished in a given period of time. The tests are accomplished by positioning a production aircraft (in this case, the second G-V to come off the assembly line) in a large test rig fitted with a network of cables, pulleys, and hydraulic mechanisms. Then, after placing hundreds of electronic theodolites all over the aircraft's structure and connecting them to a computer network in the structural engineering department, engineers could monitor the airframe's response as it was slowly flexed and vibrated to determine whether it would withstand the rigors of flight. A computer program guided and recorded the process, which seemed like a high-tech Spanish Inquisition. At no time would the computer allow anything to actually break; instead it listened to the electronic "screams" of the components as the airframe was subjected to a lifetime of abuse. The airplane would later be delivered to a happy customer.

In another building, I was shown the Integration Test Facility (ITF). The $3-million ITF comprises a full-size mock-up of the G-V's cockpit and electrical systems. This is where the G-V's numerous avionics, crew advisory system, fault prediction/trend monitoring, as well as the airplane's flight- and engine-control systems were being brought together for final installation troubleshooting (prior to production), as well as the final burn-in tests of the airplane's new avionics and engine full authority digital engine control (FADEC) system. The G-V will be the first FADEC-equipped Gulfstream, which brings the airplane up to the same safety standards as the latest heavy transports used by the airlines. Basically, FADEC provides a computer control—in the case of the G-V, dual FADEC—for the engines, making them as optimally responsive and fuel efficient as possible. It also increases the life of the powerplants and therefore reduces operating costs. The ITF cockpit is presented in high fidelity—even flight control feel is duplicated via computer simulation. The information gleaned from the ITF is dis-

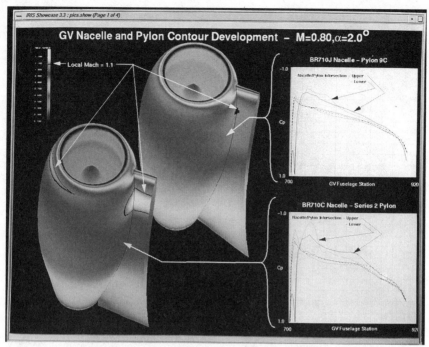

Figure 1.7   Computational fluid dynamics (CFD) programs enable engineers to visualize airflow patterns around the airframes they're developing. This figure shows a typical display from a G-V nacelle and pylon at Mach 0.80. Local Mach areas are above Mach 1 at arrows. (Gulfstream)

tributed to the flight operations departments responsible for preparing the G-V's flight, operations, and maintenance manuals.

The ITF had an informal, entrepreneurial atmosphere, but the team members were intent and all business. Masses of wire bundles were strung across scaffolding that simulated the cabin of the aircraft. Stacks of computer processors on a 6-foot-high dolly simulated the FADEC. The only thing that resembled an airplane was the exterior of the cockpit area.

Also benefiting from the ITF is FlightSafety, Inc., which contributed $500,000 to the cost of building the ITF, as well as some of the simulation effects. FlightSafety is the training organization that is building and operating the flight simulators on which G-V flight crews will train. As a part of the ITF team, FlightSafety will be able to provide high-fidelity training simulators even as the first aircraft are rolled off the assembly line, instead of having to wait until the flight test program is completed and the airplane certificated, as has been the case until now. (I talk more about flight simulation and computer-based training later in the book).

Figure 1.8 Gulfstream's Integration Test Facility (ITF) in the company's Savannah plant is used to develop to avionics and other computer-intensive systems. (Gulfstream)

Preston Henne, Gulfstream's senior vice president for the G-V program, led the G-V development effort. Most of Henne's career has been as a wing and propulsion aerodynamicist, working 25 years at McDonnell Douglas before joining Gulfstream in 1994. At McDonnell Douglas, he worked primarily on commercial aircraft development but also led the wing design team for the C-17 military transport program. This ultimately led to the development of computer code used for computational fluid dynamic programs, including the first 3-D transonic inverse method, that were later used in the development of the C-17's wing design. As many as 50 wing configurations could now be analyzed on the computer, enabling C-17 engineers to eliminate much of the time spent in the wind tunnel. "We analyzed them all in a very short period of time," said Henne, "and then picked a final configuration, tested it, confirmed that it did exactly what it was expected to do in the wind tunnel test, and we were done." But he doesn't think we've seen the last of wind tunnels. Each has its place in design activity, and each has strong points, Henne believes.

Later, Henne became the chief design engineer on the MD-80 and the MD-UHB (an experimental ultra-high-bypass, unducted fan engine

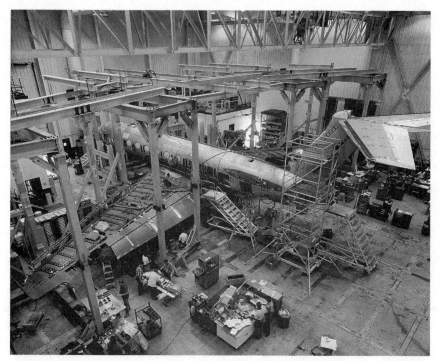

Figure 1.9 Computer systems are playing a major role in aircraft manufacturing. In this Gulfstream Structural Test Facility, a production airplane is wired to sensors connected to the company's mainframe computer to evaluate how much stress the airplane can be expected to tolerate. This is a nondestructive process; the test airplane will be sold to a customer. (Gulfstream)

design), and eventually became the program manager for the MD-90, which was certificated in 1994. That program was very satisfying to Henne, both professionally and personally. "I was responsible for the whole thing," he said. "But after the rollout, I didn't have a clue what I'd do next. Then Gulfstream gave me a call and they asked me if I could do it again."

Henne sees computational fluid dynamics as critical to the development of today's aircraft. Gulfstream uses about 20 different CFD programs that perform a number of analytical tasks, depending on the problem being solved. CFD is being taught, says Henne, in many university-level structural engineering and aerodynamics courses. Engineering department new hires at companies such as Gulfstream, Boeing, Lockheed Martin, and others, can expect to be coached through running a particular CFD code on a particular problem to gain experience with that company's version of CFD code. The biggest problem for schools, Henne believes, is keeping up with developments in engineering software. "It's not as quite as bad as the PC

Figure 1.10 The Gulfstream G-V's avionics systems were developed in the ITF. The aircraft's flight simulator programming was developed here as well, with assistance from FlightSafety International engineers. (Gulfstream)

environment, but the code you use this year will be obsolete in two or three years. It's probably tough to keep up with that on a university level."[3]

Scaled Composites, Inc.

Scaled Composites, Inc., is probably best known as the producer of some of the world's most unconventional and highly regarded aircraft. Scaled (and its sister company, Wyman Gordon Composites) is also at the forefront of moldless composite vehicle construction techniques. Instead of building hundreds or thousands of airframes as most manufacturers do, Scaled specializes in the fabrication of proof-of-concept aircraft and other aerodynamic vehicles for a wide variety of companies, including some of the world's largest OEMs.

Since its founding in 1981 by president and CEO Burt Rutan, Scaled has refined its design, fabrication, structural test, and flight test methods and services with a series of completed and continuing programs. Among these are the subscale Starship 1 proof-of-concept

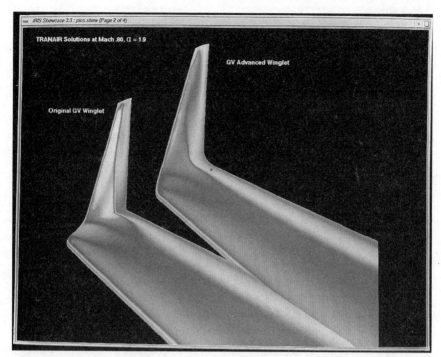

Figure 1.11   CFD was used to improve the G-V winglet design. (Gulfstream)

(POC) twin pusher turboprop airplane for Beech (now Raytheon) Aircraft; a variety of long-endurance, high-altitude reconnaissance aircraft; remotely piloted vehicles; rocket-powered vehicles and rocket booster control fins; the ill-fated Pond Racer unlimited class racing aircraft; the General Motors Ultralite technology demonstrator automobile; and the structural design, tooling, fabrication, and static-load testing of an 85-foot rigid sail/airfoil for the America's Cup Challenge Race.

The company, located about 80 miles north of Los Angeles in a 60,000-square-foot facility at Mojave Airport Civilian Flight Test Center, adjacent to the Edwards Air Force Base, is thriving. The company's test pilots and engineers maintain a busy schedule with about 360 days a year of ideal flying weather. Rutan remains actively involved in the conceptual development of new aircraft and day-to-day management of the company.

Scaled continues to break new ground in aerodynamics design applications. Recently, engineers were working on a NASA project for a high-altitude research UAV called Raptor or, officially, "D2." The airplane is in flight test. Another project, called Freewing, is another

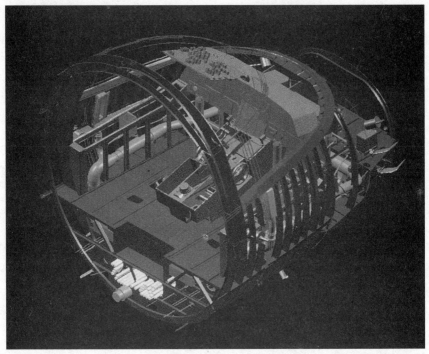

Figure 1.12   A 3-D CATIA view of the Gulfstream G-V cockpit. Each of the airplane's major subsystems is color-coded on engineers' workstations. (Gulfstream)

small (less than 300 pounds) reconnaissance UAV. Freewing is unusual in that it has a tilting fuselage and a wing that pivots about it. On take-off and landing, the fuselage tilts upward, vectoring the thrust of the nose-mounted propeller, so the aircraft "hangs on the prop" for maximum performance. Once airborne, the fuselage tilts down to align with the wings in a more conventional geometry.

Dan Cooney, chief engineer, described Scaled Composites' use of computerized engineering tools: "At our facility we stress getting low-cost prototypes into the air as rapidly as possible, so we don't spend as much time in computer analysis and simulation as some of the larger companies do. But all the basic components are still there. We do stability and control-type calculations, various types of stress work, and performance calculations [on computers]. And then when it comes time to actually make the airplane, the engineers sit down at their PCs or Macs [the company has about 25 PCs and about 50 Macintosh personal computers] and turn out drawings at a quality that can be used in the shop."

The company uses CATIA about 10 percent of the time (it has four CATIA "seats," or licensed applications), and the remaining 90 percent of the time it uses the Vellum 3D design system, created by Sunnyvale,

Figure 1.13 Computers are used on the factory floor to coordinate assembly with materials and manpower requirements. This is a view of Gulfstream's G-IV/G-V production facility in Savannah. The two aircraft share many common parts, but keeping track of them all would be nearly impossible without computer aids. (Gulfstream)

California-based Ashlar, Inc., for 3-D wire-frame modeling and 2-D drawing generation. Like many aircraft OEMs, the company also uses NASA-developed computational fluid dynamics codes on an IBM RS/6000 workstation.

Cooney said engineering students need to be computer-literate: "They need CAD skills up through wire frame. As computers get more powerful, solids and solids modeling will eventually become more important, but right now that's of very little importance to us. 3-D wire framing skills are very important, and, along with that, the creation of 2-D drafted drawings. Being able to communicate with the shop is very important, too. Our engineers also need to be good technical writers. Good technical writers are very hard to come by, and I don't think students are spending enough time developing that skill. Basic spreadsheet use is helpful, too—we use Excel and Visual Basic, for example."

Scaled Composites provides training on Ashlar Vellum, although job seekers who have hands-on experience with CATIA will stand out in front of other job candidates. The use of a drawing board is optional, said Cooney. "We have one or two more artistic people who prefer to

Figure 1.14 Honeywell technicians inspecting new flat-plate displays. The need for computer-literate people with an aviation interest is strong, say avionics managers. (Honeywell)

work on a board. If they are very creative, they're welcome to work that way."

Engineers at Scaled Composites come from a mix of traditional and nonmainstream aviation engineering backgrounds, but job applicants with only metal aircraft backgrounds "won't stand much of chance of being hired," said Cooney, since the company's aircraft are made mostly of composite materials, which call for very specific knowledge of those materials. The company hires new graduates and seasoned veterans of the engineering world, but almost all have had experience building homebuilt aircraft, and most have at least a private pilot license. "The hands-on skills to be able to construct and fly an aircraft are the skills we look for," said Cooney. "This is how we choose our people. That's probably not what you'd hear from [another manufacturer]."[4]

Fairchild Aircraft, Inc.

San Antonio-based Fairchild Aircraft, Inc., is a major designer and manufacturer of military and commercial aircraft. The company manufacturers the 19-passenger turboprop Metro 23, the C-267B military special-mission aircraft, and the multimission surveillance aircraft (MMSA). More than 950 Metros are operating worldwide in passenger and freight-carrying operations.

The company recently upgraded its CAD software by migrating from an older, mainframe-based system to more economical desktop PCs. Rather than lose more than 10,000 electronic files from its older system (commonly referred to as a *legacy system*), Fairchild sent an optical drive containing the files to an outside vendor for translation to Autodesk's AutoCAD system format. Autodesk, based in San Raphael, California, is one of the world's foremost producers of CAD/CAM systems.

Since November 1994, Fairchild has been using a variety of Autodesk programs to design new aircraft and modifications for older designs. AutoCAD is used for all drafting applications—from part and electrical wiring layouts to designing aircraft interiors.[5]

CAD for everyman

Not every aircraft manufacturer can afford a multimillion-dollar investment in high-end CAD/CAM software tools. Homebuilders have found that desktop CAD software allows them to develop innovative designs that rival those coming from the major aircraft OEMs. As testimony to that, witness the annual Experimental Aircraft Association (EAA) gathering at Wittman Field in Oshkosh, Wisconsin. Many of the aircraft that arrive there each year were originally developed using off-the-shelf, generic CAD software or shareware developed specifically for the homebuilder. The shareware is readily available on the Internet, BBSs, and online services that cater to homebuilders and engineering groups, such as CompuServe's AVSIG, Aviation Week Group Information Service, and AOPA Online forums. One can find tools that enable homebuilders to develop, for example, sophisticated airfoils using proven airfoil designs.

Magazines such as *CADalyst* and *AutoCAD* offer a wealth of advertisements for training materials and add-on (plug-in) software modules produced by third parties that add features and capabilities to generic CAD application programs.

Other Aviation Applications of CAD

Along with the revolution in traditional aircraft design, exciting new aviation-related uses of CAD are appearing. At the same time, new products are being designed to expedite and refine production and design processes without requiring OEMs to convert data files or discard their current systems.

Product data management
and ancillary software

An important aspect of designing aircraft with today's 2-D and 3-D computer tools is the management of the vast amount of information such an effort produces. When teams of engineers are using tens or even hundreds of enterprise-wide, networked computers, managing the staggering flow of information can be a challenge. This use has created a need for a new category of software called product data management (PDM) and opportunities for software systems specialists to administer and maintain the management software. In engineering and computer development circles, PDM is considered to be one of the fastest-growing areas in engineering.

The most effective PDM systems integrate a variety of engineering systems and related software. PDM systems create and manage "data about the data" and thus can track the progress of designs, revisions, and project status milestones. PDM vendors also provide modules for engineering process control and security, indexing, data reviews, and distribution in various media. Some PDM systems are provided by CAD/CAM vendors as a support product, while other companies provide PDM as a stand-alone system.

Although they are developed by software firms for a variety of industries, enterprise-wide information management systems, sometimes called enterprise resource planning (ERP) systems, are another area of rapid development in aviation and aerospace engineering. These systems provide integrated management of a company's resources and information to reduce order-fulfillment cycle times, improve operating efficiencies, and measure company performance against defined objectives. Such systems are particularly important to aviation manufacturing suppliers, which need to better manage make-to-stock and make-to-order production environments.

Itasca, Illinois-based CIMLINC, Inc.'s Manufacturing Solutions Division is typical of software vendors jumping into the aviation PDM and ERP arena. The Boeing Company recently selected CIMLINC to provide Boeing's Commercial Airplane Group with a $6.8-million system to integrate process plans (how-to-build information) and another to deliver the process plans to the manufacturing floor via several sources. Boeing expects the on-demand access to complete process plans will lead to faster and higher quality production. CIMLINC also is providing Boeing with on-site integration services to link its applications to Boeing's various ERP, CAD, and PDM systems.

CIMLINC describes its applications as "process-ware"—designed for the sole purpose of speeding and improving production and design processes without requiring manufacturers to discard their current systems or convert to new formats. In the aviation field, CIMLINC's products are also used by General Electric, Northrop Grumman, Pratt & Whitney, AlliedSignal, McDonnell Douglas, SNECMA, and Mitsubishi Industries.[6]

Numeric and symbolic math programs

Numeric and symbolic math programs, another fairly new engineering category, are used to turn a computer into a super calculator. These increasingly graphical number-crunching programs calculate statistics, calculus, differential equations, mathematical simulations, and other complex equations often called for in the creation of today's high-performance aircraft. Numeric and symbolic math programs run on computers ranging from laptop PCs to supercomputers, and the cost of the software has been plummeting in recent years. Programs that once cost tens of thousands of dollars can now be had for a few hundred dollars—a boon to engineering students.

3-D animation tools keeping skies safer

Like their partners in air safety—pilots—air traffic controllers receive both classroom and simulator training as well as regularly scheduled on-the-job instruction. A major challenge to the FAA Academy in Oklahoma City is creating classroom materials that can communicate complex materials effectively and inexpensively in the field. In recent years, the academy has been addressing these challenges by developing interactive, computer-generated 3-D animations. One of the software packages used to develop these systems are Autodesk's 3D Studios and Animator Pro.

"The FAA works diligently to stay at the leading edge in terms of operations and training," said Gwen Sawyer, manager of the Air Traffic Division's System Support Branch. "With interactive 3-D animations, we're able not only to improve the classroom instruction here at the FAA Academy, but by integrating the animations with multimedia authoring software, we can develop and deploy cost-effective computer-based instruction and facility training."

The Air Traffic Division's 135-person staff is responsible for a student population of some 22,000 controllers, administrators, and traffic managers. Classroom instruction is complemented by simulator training, also conducted at the academy, and by refresher and on-the-job education at regional facilities throughout the United States.

The academy's simulator training is multifaceted, ranging from basic simulations that provide the student a top-down view of an airport to advanced, interactive simulations that place the student in a control-tower setting. The cadre of controllers is a diverse population, so the training has to be comprehensive, including everything from specialized staff occupations to aviation fundamentals to advanced ATC radar system operation.

According to Sawyer, the academy had relied on traditional instructional materials, such as whiteboards, aircraft models, and 35mm slides. Some concepts, such as airspace structures, are particularly difficult to convey in two dimensions.

Susan Buriak, project manager for the FAA Academy's Interactive Instructional Delivery Systems (IIDS) operation, was charged with developing systems for improving instructional delivery and deploying networked computer-based training (CBT). In January 1993, she visited the U.S. Navy's Landing Signal Officer (LSO) School to learn about its computerized instructional system.

The Navy system features 3-D animations of a sea-tossed aircraft carrier deck. "I realized that these sorts of animations would be an excellent means of communicating many of the subjects we deal with at the academy, especially radar training and weather," said Buriak.

The LSO animations had been produced with Autodesk 3D Studio software. Within six months, the academy was developing curriculum content and prototype training programs with similar Autodesk products, based on its earlier 35mm-slide teaching aids.

The FAA Academy animation software is installed on 486 PCs (with Pentium systems soon to follow), some for training aid development, with one PC in each classroom as a display server. The classroom system can be controlled with an infrared remote control device. Complex animations can be produced in as little as a day.

The Academy's airspace animation is typical of what is used. The program begins with a top-down view of the Oklahoma City area and continues by showing a 3-D version of the city and the approaches to the airport. Some of the buildings, airport facilities, and aircraft were "built" from scratch with Autodesk 3D Studio's solids-modeling capabilities; other visual elements were imported as .DXF files and modified. The airspace sectors are made "transparent" and in varying colors so that animators can illustrate an aircraft transitioning from one category of airspace to the next. Similar techniques have been used at the academy to produce virtual radar scopes that can demonstrate realistic, typical radar scope scenarios. The programs are imported to Macromedia Authorware. Once in Authorware, the animation content is distributed to various regional FAA facilities.

"As proud as we are of our instructional materials," said Buriak, "we're even more satisfied with how the program contributes to the safety of the skies."[7]

Keeping track of Kalamazoo

Keeping track of the changing infrastructural elements of Kalamazoo (Michigan) Battle Creek International Airport used to be a costly, laborious, and inexact process, especially when providing information to outside contractors. Blueprints got damaged, marked up, or lost altogether.

Since 1992, however, information about the airport's infrastructure has been standardized and stored on computer in the form of AutoCAD drawings. Now, when an engineering project is required, customized drawings of the features relevant to each project can be called up, and layers can be turned on or off, according to their importance.

The AutoCAD maps have been used, for example, for documenting projects such as updating the airport's security system and the relocation of a taxiway. For both projects, the drawings were manipulated to determine the most desirable outcomes, and changes were made quickly.[8]

Electronic landlords

The Massachusetts Port Authority uses off-the-shelf CAD systems and custom-developed software to manage and track more than 11 million square feet of space and leases at Boston's Logan International Airport and Hanscom Airport in nearby Bedford, Massachusetts.

More than 23 million passengers use Logan each year, making it one of the 10 busiest airports in the United States. It's no wonder that some 350 companies operate at Logan and Hanscom, which caters to business and private/recreational flying activities. Most of these companies rent space from the Massachusetts Port Authority (Massport), the agency that also operates the port facilities within Boston harbor, on which Logan is located. Massport's Airport Leasing office is charged with keeping track of the space and leases at the two facilities.

Until recently, information on tenant leases and operating agreements was maintained in stand-alone relational databases. Graphics, in the form of leased-premises exhibits, were produced either by the air carriers or Massport's engineering department. Because these exhibits (similar to blueprints) were manually drafted and changed often, it was nearly impossible to keep the information current. Maintaining accurate billing records was often difficult. Negotiations with tenants were sometimes compromised by the time and resources required to manually develop and update the data.

In 1991, Massport began working with Autodesk to improve the Massport lease-management system, which became known as the Tenant and Property Management Information System (TAPIS). One of the largest terminals at Logan, Terminal C, was chosen as the pilot for the project. Existing manual drawings were converted into AutoCAD format and field-verified for accuracy. A new graphical interface and a relational database were developed, based on a unique set of codes for each space to be tracked. The codes were used to tie together the TAPIS textual, spatial, and graphical databases. The pilot project was completed in January 1992. Today, Massport uses the Windows-based system, running on a Novell network, to easily access data, produce lease-exhibit drawings, generate text, and provide billings for most of the two airports' tenants. It is used to track more than 1800 spaces, with more to come. With the addition of a program called CORTS (Concession Revenue Tracking System), the airport business office is using the system to track revenue and sales information from airport concessions. Other complementary programs that share data common to TAPIS are being developed or evaluated, including construction permit tracking, underground utilities systems, and an employee parking database.

These are but a few examples of the many engineering applications used in aviation. As computer capability increases, so too will the use of CAD/CAM and related systems. The use of CAD/CAM is expanding globally, especially in Russia and Ukraine, which are trying to quickly bring their aerospace industries back to life and become competitive with their Western counterparts. Indeed, some analysts believe CAD/CAM will be the key to their success.

As you can see, the field of aviation-related engineering isn't limited to creating aircraft. It presents a range of career possibilities and business opportunities for individuals who are adept with engineering skills and software.

References

1. Interviews with Boeing Airplane Company officials and corporate literature, 1995.
2. Interviews with Lockheed Martin personnel and IBM corporate literature, 1995.
3. Interviews with Gulfstream Aerospace Corp. personnel; Gulfstream Fact Book, *Setting the Standards Others Follow* (internal corporate history), no date of publication; IBM corporate literature, 1995.
4. Interview with Scaled Composites personnel and company literature, 1995.
5. Autodesk, Inc., corporate literature, 1995, supplied by Autodesk, Inc..
6. CIMLINC, Inc., corporate backgrounder downloaded from CAD Forum on CompuServe Information Service, 1995.
7. Autodesk corporate literature, 1995.
8. Ibid.

Flight-Deck Computing Systems

Some of the most profound advances in aviation computing are occurring in the cockpits of modern aircraft. While these advances promise to make flying safer and more efficient than ever, they are also redefining what a cockpit is and the nature of the work flight crews perform. Avionics advances—particularly in satellite communications—are also going to play an increasingly important role, both on the flight deck and in the passenger cabin area, as well as in air traffic management. This chapter looks at aspects of these subjects that relate to information processing in the aviation environment.

Almost overnight, the flight decks of modern aircraft have become airborne computer centers. In a span of less than 20 years, the electromechanical (EM) and vacuum-powered displays and instruments of earlier airline, military, business, and general aviation aircraft have been replaced with "suites" of integrated digital electronic sensors and displays that resemble, in form and function, advanced computer systems. Where cockpits were once a confusing array (to the untrained eye, anyway) of frequently cantankerous "steam-gauge" instruments, buttons, levers, and dials, today's cockpits increasingly resemble video arcades—complete with sound effects, motion, and flashing lights. Indeed, it probably wouldn't be all that surprising to find a place to insert a couple of quarters on the flight deck of a Boeing 777, Gulfstream G-V, or a B-2 bomber.

In the past, avionics software tended to be simple and required only a handful of programmers to complete the software aspects of a given project. Today, however, software is much more complex and can require large teams of software designers, engineers, and technicians to achieve the desired results. The development of "glass-cockpit" avionics software now constitutes between 75 and 90 percent of the cost of the systems. Some in the industry are concerned that, in many

Figure 2.1 The flight deck of the Boeing 777. The airplane's next-generation Aircraft Information Management System (AIMS) avionics computers provide the airplane with unprecedented redundancy and efficiency. Cockpit instruments are provided by active-matrix liquid crystal display (AMLCD) displays. (Honeywell)

cases, the programmers are too far removed from the flight environment to appreciate or even understand what it is they are being asked to create. Other critics assert that the aircraft manufacturers put too much faith in automation systems that might not be fully understood by the crews that use them—with sometimes tragic results.

Breakthroughs

Nevertheless, avionics development is surging along at breakneck speed, and shows no hint of letting up. Consider the following developments:

- The instrument displays on the latest civil aircraft are now being presented on 10-by-10-inch backlit liquid crystal displays (LCDs). These screens can instantly take over the display function of almost any other display unit in the cockpit. Such multifunction displays have been developed to reduce display clutter and ambiguity, and, therefore, cockpit workload. Similar developments are taking place in the cockpits of military aircraft. They are expected to have significantly longer lives than their EM counterparts, simply because they have few, if any, moving parts.

- Head-up displays (HUDs), once only found in the cockpits of advanced fighters, are now appearing in airliners and business aircraft. These remarkable systems enable pilots to keep their eyes outside the cockpit, looking for traffic, rather than focused on the instrument panel. For those unfamiliar with HUD, the difference is significant, especially when conducting an instrument approach to minimums, when making the transition from the gauges (the instrument panel) to a reliable view of the outside world (a runway environment on a dark and foggy night, for example) can take as long as 15 seconds. The use of HUD can make the transition almost imperceptible, rather than startling and possibly disorienting.

- In a growing number of military and law-enforcement aircraft, the use of advanced display technology has revolutionized the pilot helmet. Electronic displays in the helmet can provide the pilot with a wealth of visual information from aircraft-mounted sensors that can see through almost any weather, smoke, and the dark of night. In military aircraft, the pilot can evaluate a potential target and direct an arsenal of weaponry at it merely by looking at it.

- In many new aircraft, the cables and pulleys of older designs have been removed from the throttle quadrant and replaced with microprocessors to create fully automated digital electronic engine controls (FADEC) systems. FADEC provides significant dividends in engine efficiency and reliability by quickly determining and adjusting engine power settings far more precisely than flight crews can.

- Maintenance data acquisition units (MDAU), health usage monitoring systems (HUMS), and other similarly named powerplant monitoring systems are becoming commonplace tools that airlines, air cargo carriers, and other fleet operators use to oversee numerous aircraft systems for wear and tear and keep them in optimal flying condition. Similarly, helicopters are increasingly being equipped with HUMS that keep a constant watch on the engine and rotor blade parameters that are part of rotary flight.

- Fly-by-wire and, more recently, experiments in fly-by-light systems have emerged as a leading technology in modern avionics. Miles of thin copper wire or fiber-optic cables and gangs of servomotors have replaced steel cables, gearboxes, and push-rods in the dark spaces of modern aircraft, reducing weight and maintenance headaches.

- Other systems, called enhanced vision systems (EVS) and synthetic vision systems (SVS), and autonomous landing systems still in development will likely improve the odds of a safe landing even more by allowing flight crews to "see" through fog, clouds, rain, and the dark of night.

■ With the development of new concepts of communication, navigation, and ATC surveillance (CNS), impending developments in air traffic management (ATM), as well as the emergence of advanced passenger information systems, the concept of "aircraft" is quickly becoming that of a "node" in an air/ground network. Such advances are also redefining the role of the flight crew from control manipulators and navigator/engineers into information resource managers.

For the sake of simplicity, I have focused our avionics tour on the information systems side of the field as much as possible, and, for the most part, stayed away from avionics hardware engineering, as that subject is beyond the scope of this book. As a colleague recently pointed out, "If it involves manipulation of data only, or displays, usually that's software. But if it gathers data, moves control surfaces, or directly affects the flight, that's electrical servomechanical engineering."

Twenty-first-Century Avionics

Many avionics developers believe aircraft have become increasingly computerized both internally, in the traditional areas of navigation, communications, and flight controls, and externally, in the entire aircraft/ATC/surveillance/communication relationship. I agree with them.

The changes that have taken place in the past 20 years have been monumental, they believe, and are only the beginning. A number of avionics engineers have defined three factors that have made it all possible:

■ Greater processing power provided by data bussing

■ Introduction of new display technologies

■ Miniaturization of avionics components

One of the people in the avionics field that I interviewed while researching this book was Gene Schwarting, director of product and strategic marketing for Collins General Aviation division of Rockwell International. He was particularly expansive on the subject and represents the basic views of many in the industry I have spoken to recently. I asked him about the transition from electromechanical cockpit displays to the dazzling digital avionics suites found on many modern aircraft.

"Inside the aircraft," said Schwarting, "the transition to a computer environment has been going on for quite a while. I suspect it started on the military side sometime in the 1980s, with the introduction of [data] bussing—with Mil Spec 1553 and that sort of thing."

Mil Spec, or more properly, MIL-STD, is the U.S. military's definition of a standard set of form, fit, and function parameters that defense contractors frequently adhere to when developing advanced weapons systems, aircraft, avionics, wiring systems, software, submarines, soldier's helmets, and so forth.

Data bussing refers to the use of common electrical pathways and data transfer schemes in computer circuits. MIL-STD 1553 is a standard that avionics developers use to digitally unite various aircraft subsystems or functions to save space and weigh and reduce avionics complexity and costs. (See the section on MIL-STD 1553 later in this chapter.)

The bussing concept is so elegantly simple that it is amazing it took until recently to develop. At its simplest level of definition, bussing allows a network of instrument and sensor subsystems to share space on a precisely timed electrical circuit in an aircraft's avionics network. Each subsystem is assigned a digital "seat" on the bus. If subsystem A has a package of information (airspeed, distance from a navaid, fuel quantity, or heading, for example) to relay to other instruments in the network, it puts that package on the seat and away goes the bus until subsystem B, needing that particular type of information, finds it sitting on the incoming bus. The package is removed and newly processed information from subsystem B is added to other seats on the bus. Then the bus continues on its circuit around the cockpit, with each subsystem adding or deleting information along the way. This process takes only milliseconds to complete (typically measured in megahertz). Data bussing techniques of different schemes are used in virtually all modern computer circuitry.

Take the Bus

Commercial aircraft avionics and airframe manufacturers began to adopt the digital bussing interface concept in the late 1970s and early 1980s. Before industry-wide standards emerged, however, Collins Avionics (now a part of Rockwell International) developed its proprietary Commercial Standard Digital Bus standard (CSDB) for use on all its newest avionics so they could "talk" to each other. Similarly, Honeywell Avionics, another major avionics company, developed Avionics Standard Communications Bus (ASCB). However, operators of aircraft that used avionics from both manufacturers in their fleets soon discovered the two bussing languages were incompatible.

To circumvent this impasse, which resembled the Beta-versus-VHS and Apple-versus-DOS consumer marketplace standoffs of the same era, each vendor quickly began to provide interfaces that made it possible for the two formats to work in concert. Meanwhile, ARINC, the airline-owned creator of air carrier industry standards, developed its own avionics bussing language, ARINC 429.

ARINC 429 and other task-specific (i.e., satcom, air data, navigation, weather radar) ARINC standards have for the most part replaced the ASCB and CSDB digital bussing formats (and ARINC 429 is giving way to ARINC 629, which is used on the Boeing 777, among other aircraft). Initially, however, ARINC 429 was slow to be adopted by the industry because few operators wanted to absorb the expense of transitioning to new avionics that were based on the ARINC 429 bussing standard. Today, however, ARINC standards are nearly universal.

"All of sudden—by the mid-1980s—everything was digital," said Schwarting. "And you were really beginning to form networks in the aircraft in the sense that everything was connected digitally. But in my view, the real events that took place to form a network concept—and I think we were the first to do this—we put the boxes into a cabinet. We put a lot of the functions into a cabinet. We called it an IAPS, for integrated avionics processing system." (Honeywell calls their cabinet equivalent IAC, for integrated avionics computer.)

"You could now share a number of the resources inside this cabinet—input/output and power supplies, for example," said Schwarting. That development, which Collins and Honeywell each experienced nearly simultaneously, allowed the avionics cabinet to become a major node in a system. Processing that was taking place inside the avionics cabinet was interconnected to the displays and controls on the flight deck. The avionics cabinet was also connected to the aircraft's navigation and air data sensors, as well as to the aircraft's control surfaces. The airborne computer center was born.

Virtually all digital bus avionics architectures are now identical, except for minor differences in functional packaging.

The Digital Aircraft Takes Flight

The military led the development of advanced avionics for defense applications, originally with the introduction of the transistor in the 1950s and 1960s. By the early 1970s, most advanced military fighter and bomber cockpits had become at least partially digital, although they were a far cry from today's standards. "Back then 'digital' meant you had shift registers, and they took up a lot of space," said Schwarting. Mission programming systems relied on bulky tape mechanisms for input, and displays tended to be limited to light-emitting diode (LED) displays for nav/com, fuel management, electronic warfare (E/W) and weapons stores display units, monochrome CRT radar displays, and alphanumeric input/output devices.

Among the first digital civil aircraft systems were found on the McDonnell Douglas DC-10s that went into service in 1972. These three-engine, wide-body transports used the most sophisticated computer applications aboard an airplane up to that point. Collins provided a flight management computer (FMC) system for the airplane that comprised a monochrome CRT display and a full alphanumeric keyboard. However, this was in the days before the introduction of CRT-based electronic flight instrument system (EFIS) displays, so the flight crew's instruments were still electromechanical flight directors (FDs) and horizontal situation displays (HSIs).

The first FMCs (also called navigation management computers, or NMCs) made it possible to electronically store and access navigation databases—information describing the fixed points in space that define airways and airway intersections (usually defined in terms of distance from a navaid), navaid and airport locations and identification, altitude restrictions, and other information required to fix the position of an aircraft at any given time. Soon, FMCs and NMCs were connected to avionics busses and were elevated to "integrated flight control systems"—smart autopilots that could be programmed to fly specific routes, altitudes, and airspeeds and, at the appropriate time, turn control over to the autopilot when the aircraft was approaching the destination airport's landing aids, such as an ILS. FMCs later developed additional capabilities, such as vertical nav (VNAV), which made it possible for the computer to comply with the limitations of published standard instrument departures (SIDs) and standard terminal arrival routes (STARs), and the ability to fly various types of holding patterns.

Later in the 1970s, the first true avionics microprocessors appeared. For Collins, the first application was the digital readout device (IND-40) used for the company's DME-40 distance-measuring equipment system. "After 1977, it was like an explosion of microprocessor applications," said Schwarting. "The processing capability just seemed to mushroom. The way that I remember that was the use of what is now called digital signal processing [DSP]. That was a whole new way of designing an RF box—a radio. The first DSP application for us was turbulence weather radar. That operates on very low transmit power—around 25 watts—and so the signal returns are even lower. The only way that thing can work is through correlation technique—you're digging these tiny little returns out of [the natural background radio] noise. We developed a special DSP chip for this that did all the processing in that thing.

"Within a year or two we were using DSP in GPS [global positioning system] receivers and, later, in satcom antennas. And now we're using it in windshear detection systems. I consider all of this to be microprocessor technology."

The digital applications curve is becoming increasingly steeper. Schwarting said that today, nearly three-quarters of all the design that goes into a new avionics system is spent on software, compared to about 10 percent in the early 1980s. The advent of the microprocessor, with the rapid increases in its power and speed, has been crucial in the development of digital avionics. "But there are a couple of other things that I put in very close proximity to that," said Schwarting. "One of them is display technology. Back in the 1970s, the ability to put a CRT into a cockpit had a lot to do with being able to make use of all of this processing power. Now you could use symbology to do a lot of the work, but you needed some type of writable display—the CRT allowed us to do that. That also began to change the way the flight crew interfaced with all the systems. It used to be all mechanical—knobs and dials. Now you could visualize things. You could look at weather radar in color, for example."

The second major advance in cockpit technology is the switch from electromechanical displays to CRT (video) and, more recently, active-matrix liquid crystal (AMLCD) flat-plate displays. Flat-plate displays first appeared in civil cockpits in the form of instantaneous vertical speed indicator (IVSI) displays associated with the Traffic Alerting and Collision Avoidance System (TCAS) in the early 1990s. They are now used for primary flight displays (PFDs), backup emergency instruments, and cabin information/entertainment monitors.

The third key element in the computerization of the cockpit has been the increasing miniaturization of avionics components and packaging. Advances such as surface mount technology have enabled avionics engineers to pack increasing amounts of capability into ever-smaller processors, which require less and less space in the aircraft. Indeed, if you were to open the cases of many avionics systems, you would find them filled with a lot of air. It is simply easier to shrink the interior of avionics components than it is to reconfigure the avionics bay of an older aircraft design.

AMLCD

The latest advance in display technology has been the introduction of flat-plate active-matrix liquid crystal diode (AMLCD) displays. Although CRT displays can provide wonderfully detailed images, they are heavy and produce a lot of heat. Attempts to squash the picture tube were either too complex (and therefore costly to produce) or provided inferior pictures. Flat-plate displays have been around in various forms since the late 1960s, but only since the mid-1980s have manufacturers (mostly in Japan) developed technologies that provided CRT-equivalent image quality and commercially viable techniques for making them in large quantities. AMLCD displays use thousands of tiny transistors embedded in the back of the screen to produce images by turning on or off associated liquid crystal cells. When the transistors are turned off, they maintain their last color setting until the screen's next refresh cycle.

In recent years, the cost of flat-plate displays has decreased sharply as a growing number of manufacturers entered this very competitive field, but AMLCD manufacturing is considered to be a multibillion-dollar industry. Experts in the field describe AMLCD uses as nearly endless and predict they will soon replace CRTs in virtually all applications.

As costs continue to plummet and the technology advances even further, AMLCD will likely take on new roles, such as Dick Tracy–style wrist-TVs, wearable computers, dashboard navigation screens, electronic books, notebooks and electronic clipboards, and wall-size monitors that can be used to display art. (It's rumored that Bill Gates already has some of these in his home).

Airborne networks

"So all of these advances, taken together, spurred on the concept of bussing and redistribution of functions and networking," said Schwarting. "They all evolved simultaneously. It was really stimulated by processing power, redistribution of processing, and the need to interconnect things in different ways. This has been going on *inside* the aircraft. My feeling is—and I don't think I'm alone on this—what you're going to see is the airplane becoming just a node on a network.

"Information will be flowing to and from the airplane in real time both for flight-guidance applications and passenger applications. Data link will make this possible. You will no longer have to carry onboard all the information you need to complete a flight. I think the network is extending outside the airplane now, and the airplane simply becomes a node on this vast communications network."

Others in the avionics industry have echoed Schwarting's comments and see this development as the most significant and far-reaching one in aviation in recent memory.

The passenger cabin is the next area due for major development, Schwarting believes. Reliable, cabin-wide local area networks and seamless communications with ground-based services are probably going to be the first area of success, which Schwarting believes will take place in the next five years. "The big issue is bandwidth. Solve that and you've solved the problem."

As we'll see later in this chapter, and elsewhere in this book, this development is closer than we might think.

Functional rethinking

"We in the avionics business," Schwarting continued, "have a hard time talking about the old discrete functions anymore—autopilots, flight management, and display processing systems—because they're not in the old discrete boxes they used to be in. Nearly everything's blurred now. The old categories of EFIS, autopilots, and radios are gone. They have been replaced by a new concept of interconnectivity."

Avionics engineers have been redefining how they think about avionics—about the "old" categories of avionics, such as navigation receivers, communications radios, radar, autopilots, powerplant indications, heating and cooling systems, hydraulics, and so on. Engineers have come up with entirely new concepts of avionics functional packaging. New or revised concepts for today's avionics include integrated flight control (which includes the traditional autopilot functions, but now also includes navigation management and related displays); integrated communications systems (the traditional RF function plus automatic dependent surveillance, or ADS), Satcom, and high frequency (HF) radios; engine management (which only recently became known as engine indicating and crew alerting systems, or EICAS); utility system management (which can include fire control, hydraulic, electrical power, and air conditioning systems); and the maintenance management system, which touches all the systems on the aircraft. "We're using this approach to rethink the ways we go about designing systems," said Schwarting.

Milestones

Collins' latest line of avionics—Pro Line 21—incorporates the new thinking about avionics functions and typifies some of the advanced products being created by leading avionics manufacturers. The baseline Pro Line 21 system uses four 10-by-8-inch AMLCD adaptive flight displays to tailor the aircraft's flight deck to the mission of the aircraft, but more or fewer displays can be used. More and different types of information can be presented on fewer displays than was previously possible. All the information required to fly the aircraft, including attitude, altitude, heading, airspeed, engine data, navigation maps, TCAS, and radar can be positioned on a single display to reduce the pilot's instrument scan. As new avionics capabilities and ADS/CNS infrastructures are added, such as ground collision avoidance systems, terrain mapping, taxi charts, and designated data displays, the system can be enhanced. The first aircraft to use the Pro Line 21 will be the Raytheon Premiere I business aircraft, which is expected to be certificated in 1997.

The new Collins system will also benefit human factors, said Schwarting. "Now, you'll no longer be forced to follow needles. You can visualize and fly flight plans that we probably haven't even defined yet. So you need to have display media to match this. The new flat-panel displays will help make this happen. You can put graphics up that will just blow your mind."

Collins engineers view Pro Line 21 as an evolutionary system, not an event in itself. The first step in this evolution was the introduction of the new display technologies. The avionics architecture is likely to change as it evolves, said Schwarting, allowing Collins engineers to continually rethink how pilots fly and behave in the new data-link, Free Flight (both of these concepts are discussed elsewhere in this book) airspace environment.

Schwarting had praise for the Honeywell/Boeing development team that produced the avionics and related systems on the Boeing 777. "That is a computer implementation that has never been done before—at least in the civil world. It appears to be well done.

"The 777 is a shared processing capability. The reason I bring up the 777 is because the verdict is not in yet on what direction processing will take in the airborne environment. [The 777] is one direction it could take—you have basically one huge processor, software partitioned to perform multiple simultaneous tasks in a shared environment. By contrast, you can have a distributed type of processing, in

Figure 2.2 Honeywell avionics development engineers test and validate cockpit avionics for the Boeing 777 at Honeywell's 777 integration validation facility in Phoenix. (Honeywell)

which you maybe have two major processing nodes—one in the display system and there's one in the flight control/flight management system which communicate."

Boeing 777 avionics program managers stressed to me that the aircraft's computer system—called AIMS (Aircraft Information Management System) is actually comprised of two major processors, and each is designed to be fully redundant.

Industry Needs

As for those entering secondary schools or the job market and considering the aerospace avionics field, Schwarting believes an interest in flying helps. "Without a doubt. It may not be a prerequisite [to have a pilot's certificate], but it would certainly would give somebody a step up over their competitors. The reason I say that is because we are going through a transition in the avionics design world where the people who are putting in seventy-five percent of the total new avionics development effort really—for the most part—don't understand the functions they are implementing. They are software writers and verifiers.

"There is almost an unnatural act that has to take place—some poor soul has to capture all the [new avionics'] requirements and translate them so that hundreds of software people can write the code. If you go back to the 1970s, the person who designed the box understood how it was going to be used. There was a lot of efficiency and finesse that went into it. You don't necessarily have to know how to fly the airplane, but you have to know the nuances of flying."

Schwarting thinks that someone who has good simulator skills or even a talent for using a desktop flight simulator program will bring a lot of value to the avionics development profession. "The software you can buy today is a good start for it," he said.

Managers at a number of leading avionics and airframe manufacturers I talked with agree that the avionics industry might be in need of a new approach to training software engineers. Schwarting's comments were typical, however.

"I don't know how to do what I'm saying here," said Schwarting, "but I'm just saying that the people who are responsible for code preparation could be more efficient, and it could be done more quickly if there was a way that they could appreciate what the impact was of what they're trying to achieve.

"I think the University of North Dakota Aerospace has some outstanding training programs. And it is that type of school that could really help someone who is interested in the industry to get some background. For a number of years we were sending some of our folks up there—engineers who didn't have a flying background—for one-

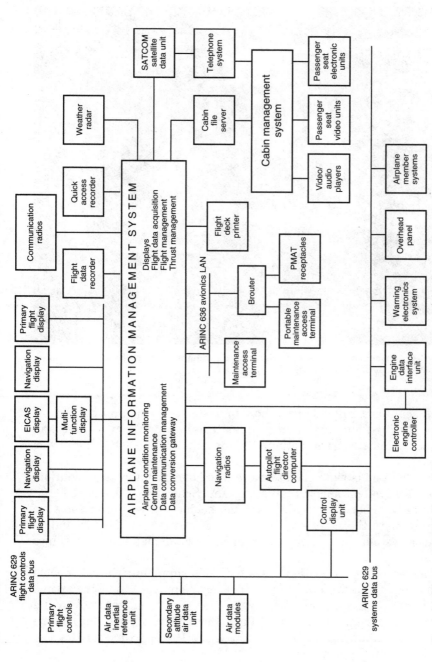

Figure 2.3 A simplified functional block diagram of the Boeing 777 Airplane Information Management System (AIMS) shows it to be a network of networks. The system comprises some 2 million lines of software code and is considered a major achievement in systems integration. (Boeing)

41

Figure 2.4 A Honeywell software engineer checks the installation of an AIMS avionics cabinet in one of the early Boeing 777s. (Honeywell)

week seminars. They learned how to design systems and how to get them certified and sat in on a radar training course and other programs. I'm sure there are others that do similar things, but theirs was great."

But the field is wide open to those who have an interest in aviation. "There's a systems aspect to this, too," said Schwarting. "We have found that the number of qualified systems people—these aren't software programmers, these are *systems* people—is way too small for the industry. The systems field is another that has high demand in the avionics world."

The secondary educational system isn't supplying enough people with adequate backgrounds in systems development, said Schwarting. "It's better than it was just a few years ago, but it's one that deserves a lot of attention—it's an emerging field. The importance of it has become a lot more significant in the last few years. It takes time for the academic community to catch up."

Another concern of Schwarting's is the FAA's inability to keep its avionics standards up to date with what is being developed by the avionics community. Schwarting thinks the FAA would be the first to admit it. The agency, he believes, just doesn't have the staff to keep up with the certifications of some of the new avionics systems in the United States or abroad. It has led to a certification backlog that began in 1994. "They have a bunch of GPS approach procedures just waiting to be certified that they can't keep up with. But I think they understand the problem, too." (There is an FAA initiative to certify about 400 GPS Category I approaches per year starting in 1997. Industry has asked FAA to ramp up to 1400 per year, but funding issues might cause a delay in the program.)

One solution might come in the form of object-oriented programming software (OOPS). Schwarting said the use of certifiable reusable OOPS is certain to become a major element of avionics development as avionics vendors look for ways to reduce the cost of creating their new products. "If you can certify an avionics software object, you'll really simplify the certification process," he said.

Avionics pace of development

"You know, this is the most exciting time to be in this business," said Schwarting. "Everything is coming to pass that we've been dreaming about for years. It's so fascinating to see all this stuff unfold. There's hardly a day goes by around here that someone doesn't come by my office and tell me, 'Gene, you gotta come see this!' "

Schwarting is still surprised at the pace of development in digital avionics, even though he is close to the action. "I am amazed at the speed with which this has taken over almost every function of an airplane. I can't imagine what the next ten years are going to be like."

Human factors and automation

Human factors benefits are already accruing from the development of today's new avionics. In aviation, human factors (HF) is the science of matching the machine to the needs and mission of the flight crew. HF is a new and vibrant area of research and development. Earlier in this chapter you saw how electromechanical (EM) instruments were being replaced by CRT and LCD-based cockpit displays. At first, most of the information being presented on electronic attitude direction indicators (EADIs) and electronic horizontal situation indicators (EHSIs) was nearly identical to the older EM displays they had replaced. But as avionics engineers developed new integration capabilities, and the need to reduce costs and system complexity grew, engineers were faced with the task of displaying a greater amount of information on smaller displays. This challenge quickly led to the introduction of the use of vertical tape displays of airspeed and altitude in the margins of the EADI. Some pilots liked them, but many did not (including me).

Figure 2.5   The Beechcraft Starship was the first civil aircraft to feature all-digital avionics. Contrast this integrated avionics suite with the cockpit of a late 1940s-era twin in Figure 2.6. (Beechcraft)

The problem, as many saw it, was that the new digitized displays didn't provide the "trend cues" that the old steam-gauge airspeed and altimeter dials did. Recent studies by the U.S. Air Force[1] have demonstrated that most pilots do indeed fly with greater precision and smoothness when presented with analog rather than digital information, so round dials are returning to many of the newest displays. Another reason for the return to some round-dial instrument configuration is the size of the latest displays, which are now commonly called primary flight displays, or PFDs (combining the function of the EADI and the EHSI), and the multifunction display (MFD), which is used to display engine indication and crew advisory system (EICAS) information and radio management displays, as well as serve as a backup if any of the PFDs fail.

At the same time, the cost of manufacturing AMLCDs is declining, so ever-larger displays are finding their way into the newest cockpits. If cost and cockpit real estate are no longer major issues, might there be a limit to the size of the displays? At this point it is anyone's guess. There is probably a point of diminishing returns, where it doesn't make

Figure 2.6 A typical predigital cockpit of a light piston twin. (B/CA)

sense to make a PFD or MFD any larger than a desktop computer monitor. Of more concern to some is whether the traditional symbols are most effective for displaying what is actually happening to the aircraft. NASA and private-sector engineering psychologists have proposed the use of a "tunnel" visual metaphor and directional vector symbols as more effective cues. These are areas being explored for military applications, but if they are found to be worthwhile, they will likely transfer to the civil avionics marketplace.

The subject of information overload in the cockpit is also being studied throughout the industry, but as yet there doesn't seem to be much of a consensus on how much information is too much. A related concern is the degree to which the cockpits of many aircraft have become automated environments. Many people in the industry are concerned that there don't appear to be consistent standards for automation—or automation training—from aircraft to aircraft and even within a single type of aircraft. NASA and other research organizations are examining this issue closely, too. Issues that are being debated include

- At what level should avionics certification agencies become involved?
- What automation standards need to be developed for the entire industry and which need only apply to specific aircraft?

■ What is the role of training in automation? Should pilots who are receiving initial and recurrent training spend more time on mastering an aircraft's normal automation procedures or with the exceptions that might or might not ever occur?

■ What types of navigation displays are appropriate in those parts of the world where navigation information might be inaccurate? It has been proposed that one pilot fly using "raw data" navigation displays while the other uses "blended" information (position computed from a variety of sources). In some cases, switching from one display to another can produce dramatic map shifts that can lead to disorientation and errors.

The Evolution of Standard Avionics Software

Today's avionics are remarkable achievements in software and hardware engineering. But they have only come about after years of effort. Among the hurdles that needed to be crossed were (1) developing a software language that would be universally accepted; (2) developing high-speed microprocessor chips on which to run the software; and (3) developing suitable, rugged bus architecture to keep the system compact, lightweight, easy to modify for various peripherals, and easy to maintain.

Ada

The first advances toward software-based avionics occurred in the 1970s. A profusion of software codes (some estimates put the figure at around 400) were beginning to be used to develop military systems. The U.S. Department of Defense (DOD) became concerned that a "Tower of Babel" situation would quickly develop, in which each systems vendor would be creating its products in its own software language and competing manufacturers' products would be incompatible. Thus, software language standardization was desirable.

The existing languages were studied to determine if a common standard could be created from them for highly complex systems that require real-time processing. None of the languages met all of the DOD requirements, so an effort began to define a new avionics software language. The final specification was published in 1980.

What emerged from the specification was the Ada programming language. The American National Standards Institute (ANSI) accepted the language in 1983, and since then, the DOD requires that all military systems be designed with Ada, also known as MIL-STD-1815.

Ada was named after Augusta Ada Byron, daughter of Lord Byron and friend of Charles Babbage, the English mathematician and inventor. Babbage is credited with developing the idea for the first mechanical computing machine, called the "analytical engine." Ada is reputed to have programmed Babbage's computer.

Ada is a structured programming language that uses modules that can be compiled separately. It is considered a very large language; compilers can require several hundreds of thousands of lines of code. Ada has been criticized for trying to be good at all tasks, and its size is an invitation to disaster, because errors can go undetected in the testing process.

Modular software

Ada has achieved one of its primary goals. Avionics software engineers can now develop products using software modules, instead of having to "reinvent the wheel" every time a new product is defined. Software modules are capable of functioning independently to create a desired result and can be created to be more or less interchangeable, much as hardware is interchangeable. Software modules can be created to accept and produce specific input and output format. Modules can then be integrated with other modules to produce a set of desired results. When a large program is broken down into modules, errors can be more easily detected and traced to their origin for debugging. This process is especially helpful when creating complex systems such as avionics suites and missile guidance systems, where teams of software programmers are working on a single project.

There are a number of ways to create software modules. The most popular design approach is called object-oriented programming (OOP). This technique involves breaking a process down into a number of components called objects. Modules are then created using the objects as focal points. Another recent software engineering development, called computer-aided software development (CASE), has made modular software development even more efficient and promises to simplify the process of simulating, testing, documenting, and managing software.

Avionics microprocessors

Until very recently, the evolutions of computer chips for military and commercial applications generally have followed parallel, but very distinct, paths. Because military airborne computer processing power requirements often preceded commercial needs, the early days of military computers were characterized by very different machines than those found in data-processing departments. Most of those machines (like the AN/AYK-14 family of computers) were relatively bulky and suitable only for larger aircraft. Even the programming languages used in the military differed significantly. The FORTRAN- and COBOL-based systems typically found in civil applications were set aside for languages like JOVIAL and CMS-2 and a wide variety of processor-peculiar assembler languages.

In the 1970s, the appearance of microprocessors like Motorola's 6800, Intel's 8080 and Rockwell's 6502, and later 16- and 32-bit progeny, blurred the differences and allowed computers to migrate into smaller aircraft. In the 1980s, the DOD tried to leap ahead of the swarm of different types of military processors and computer languages by developing what it believed would be a tri-service common computer processor architecture (known as MIL-STD-1750) and high-level programming language (MIL-STD-1815—Ada, discussed earlier) which would take the military well into the twenty-first century.

Although Ada has gained wide acceptance (it's entering its second generation as a standard, has been adopted as a civil aeronautical standard in the form of ARINC 613, and was used as the main programming language for the Boeing 777 AIMS computers), the same cannot be said of MIL-STD-1750.

By the time it started entering operational use, the 16-bit architecture of MIL-STD-1750 had been surpassed by the likes of SUN Microsystems SPARC architecture, Motorola's 680X0 and Intel's 80×86 families of microprocessors. The high cost of military computers, the withdrawal of many integrated circuit (IC) manufacturers from the Mil-Spec microcircuit business, the increasing reliability and ruggedness of the less expensive and widely available commercial counterparts, and shrinking defense budgets are making way for the prevalent use of "ruggedized" commercial computers and commercial standards and specifications in airborne applications.

MIL-STD-1553

Although developed for the DOD in the early 1970s, MIL-STD-1553B is still very much alive in the 1990s and will be with us for quite a while, say avionics engineers. Almost every military aircraft built after the B-1 bomber and the F-15 and F-18 fighters have one or more 1553 data busses on board. There is a weapon data bus version of 1553, which is part of MIL-STD-1760, a NATO (European) version defined by STANAG 3838, and a fiber-optic version of 1553 known as 1773. The F-22 fighter and RAH-66 attack helicopter, currently under development, are slated to use a faster data bus known as the high-speed data bus (HSDB). However, there will still be some 1553 busses onboard to accept older (today's) avionics.

The 1553 bus has been used not only in aircraft, but also in satellites (satellite manufacturers now prefer the fiber-optic 1773 standard), tanks, factories, and nuclear plants because of its relatively good immunity to electromagnetic noise. There is an ongoing effort to return 1553 to its commercial roots with the Society of Automotive Engineers (SAE), where it would become known as SAE AS 15531.

The 1553 data bus concept enables multiple "black boxes" to send and receive data serially over a shielded twisted-pair cable. As many as 32 devices can be connected to a single 1553 data bus. All data transfers are executed under the control of one of the black boxes, which is designated the bus controller (the need for this bus controller is seen as one of its main weaknesses, however). All other black boxes in this type of system are known as remote terminals. There is another type of 1553 device known as a bus monitor, which only listens to communications occurring on the data bus. Data is transferred using a clock rate of 1 MHz. Once the 1553 command overhead is taken out, a 1553 data bus allows a maximum data transfer rate of 800 kilobytes per second (Kbps). (Although there are several variables to consider when measuring a computer's speed, the clock rate is the speed at which operations proceed within the computer's internal memory. By comparison, the original IBM 8088 processor has a clock speed of 4.77 MHz, while the original IBM PC 80286 chip has a speed of 6 MHz.)

MIL-STD-1553 is in no way similar to ARINC 429, which is unidirectional (one transmitter and as many as 20 receivers). Also, in 1553, the definition of data formats is left to each avionics manufacturer. It also bears little similarity to the newer ARINC 629 standard that is now used on the Boeing 777. The data bus that most closely resembles 1553 is the ASCB, developed by Honeywell.

RTCA DO-160c

RTCA, Inc., is an association of aeronautical associations of the United States from both government and industry. It is dedicated to the advancement of aeronautics. RTCA seeks sound technical solutions to problems involving the application of electronics and telecommunications to aeronautical operations. Its objective is the resolution of such issues through consensus. The findings of RTCA are usually recommendations made to all organizations concerned. Since RTCA is not an official agency of the United States government, its recommendations shouldn't be regarded as statements of official government policy; however, when the FAA develops new avionics certification requirements, it leans heavily on RTCA recommendations.

The FAA, Joint Airworthiness Authority (JAA), and other civil aviation authorities will not certificate electronic equipment until it has been demonstrated to comply with RTCA DO-160c requirements. The military equivalent of this document is a combination of MIL-STD-810 (environmental conditions), and MIL-STD-461 (electromagnetic interference and compatibility) requirements.

Avionics storage media

In high-speed, high-density applications, such as would be found on the flight deck or an avionics bay, cooling is one of the toughest issues airborne computers have to deal with. The flight computers found on the F-22 advanced fighter and RAH-66 helicopter require a sophisticated liquid cooling system. Otherwise, the biggest developmental hurdle for aircraft computer makers is the ability to operate normally in the demanding environment of an aircraft (changing altitude, temperature, and humidity, sand and dust, salt spray, aviation fluids, fungus, etc.).

In the military, the pilot interface to the flight management computer/navigation management computer (FMC/NMC) is generally referred to as a control display unit (CDU) or control and navigation display unit (CNDU). Early systems often required a bulky magnetic tape device sometimes called a tape transport cartridge. A few avionics systems required that a ground crew load the data into the aircraft's mission computer using a memory loader/verifier.

In the 1980s, the military experimented with many forms of media, including electronically erasable programmable read-only memory (EEPROM) and bubble storage devices, to name only two. Increasingly, removable hard drives, flash ROM cartridges, and the standard 3.5-inch floppy disk are being accepted as the usual means of storing navigation and flight-planning databases.

New technologies for storing and retrieving large amounts of information in the often inhospitable environment are being explored and promise to transform the way we manage information in the aviation industry, both during flight and on the ground. On next-generation aircraft, it will be possible to store an entire set of instrument approach plates, as well as complete flight, operations, and maintenance manuals, on a CD you can slip into your pocket. Approach chart and flight manual revisions, once the bane of first officers, will be a thing of the past. Although there have been attempts to develop electronic versions of approach charts in the past, none have been completely satisfactory.

An airborne electronic library system (ELS), such as that used on the Boeing 777, can become a primary element of the aircraft. On the aircraft an ELS can be linked to navigation displays (airway, sectional, and instrument approach charts), EICAS (checklists and flight manuals) and FMS displays (aircraft performance data), and via air/ground data links to ATC, air carrier, or flight department operations or maintenance facilities.

Gulfstream G-V's digital avionics. The latest developments in advanced digital avionics often appear on business aircraft, as these aircraft are often regarded the ultimate in operational flexibility and safety by their operators. The Gulfstream Aerospace G-V business jet is a recent example of an aircraft that is taking advantage of the digital revolution in avionics.

The cockpit and cabin of this airplane, when certificated later in 1996, will feature many of the advanced avionics and communications systems available today. Indeed, many of the G-V systems appeared in the company's G-IVB model, which was one of the first business aircraft to feature an all-glass cockpit.

"All-glass" denotes the use of advanced display systems that rely on CRT or LCD instrument panels instead of electromechanical instruments. (It should be noted, however, that you'll still find a few EM instruments, such as attitude and airspeed indicators, in most all-glass cockpits, as insurance should the airplane ever experience a total display failure, but even these backup instruments are beginning to be replaced with independently powered digital displays.)

The G-V cockpit features a suite of Honeywell SPZ 8500 avionics that uses 8-by-8-inch LCD primary flight and multifunction displays. The G-V will also feature full vertical and lateral nav (VNAV/LNAV) and auto-throttle performance from takeoff to touchdown. Autothrottles provide an added measure of precision and efficiency over the engine power settings.

Upon certification, the airplane will be offered with an optional Honeywell/GEC Marconi head-up display (HUD), a new synoptic crew advisory system (CAS), and an enhanced digital ground proximity warning system (GPWS) with a digital terrain database.

Gulfstream wants to make the airplane as versatile as possible, so it can eventually be approved for near zero-zero landing capability by means of a forward-looking infrared radiometer-based (FLIR) enhanced vision system (EVS).

The airplane will also be available with a customer-configurable, digital maintenance data acquisition unit (MDAU) that is similar to the maintenance computers found on the Boeing 777 and 747-400 and McDonnell Douglas MD-11. The off-the-shelf design MDAU was initially installed as part of the G-V's full-authority digital engine control (FADEC) system, but its role has been expanded. Both while in flight and on the ground, the information from the MDAU is compared to the airplane's master minimum equipment list (MMEL) so it can facilitate dispatch reliability while controlling maintenance costs. For example, if a maintenance parameter (such as an engine temperature or pressure limit) is exceeded, the MDAU logs it and forwards a message to the crew advisory system. Upon arrival at the airplane's maintenance base, the MDAU's memory can be downloaded for review by mainte-nance personnel, and any appropriate measures can be taken. FADEC provides extra engine reliability by automatically setting the power to prevent engine overspeed or overboosting conditions.

The Starship arrives—and departs.   One of the first—and certainly one of the most striking—aircraft to employ advanced avionics technology as a major feature of virtually all of its flight and nav management systems was the Beechcraft Starship. The company (now owned by Raytheon) began development of the Starship in secrecy in 1979. Beechcraft spent some $300 million in technology development and certification on the Starship. When it was introduced at the 1983 annual National Busi-ness Aircraft Association (NBAA) convention, this all-composite, twin-turboprop canard-wing business aircraft was seen as a major milestone in advanced technology manufacturing and featured a suite of what later became known as Collins' Pro Line 4 line of all-digital avionics.

The airplane's cockpit featured 14 CRT displays (12 color and two monochrome). The largest CRTs were two primary flight displays (PFDs) and two primary navigation displays (PND)—one each for the pilot and copilot.

The PFD displayed an electronic version of a flight director/attitude indicator, while the PND provided a bird's-eye view of the aircraft's heading and track in relation to navaids along the route of flight. Nearby, the airspeed indicator and altimeter/vertical speed indicator were electronic versions of their round, electromechanical predeces-sors. The airspeed indicator display included a computer-generated true airspeed readout, and the airspeed needles indicated airspeed trend-vector displays—helpful touches that were previously unheard of in all but a handful of aircraft. A pair of multifunction displays (MFDs) in the center of the console provided flight crews with EICAS displays—engine power, temperature, and fuel parameters. The MFDs also provided graphical displays of other aircraft system status indica-tions that were previously displayed on a master warning/caution panel on the glare shields of most turbine aircraft, although some of these displays could be found in their traditional locations. Many of the Starship's displays were duplicated in digital as well as steam-gauge graphical formats. Although it might sound like a confusing array of information, I quickly found the Starship cockpit a user-friendly work-place, and the fine handling qualities of the airplane made it a pleasure to fly with a minimum of introduction.

Unfortunately, despite its impressive handling characteristics and dazzling features, the dramatically sleek-looking airplane (originally designed by Scaled Composites) proved hard to sell. Compared to its more traditional competitors, the aircraft was deemed relatively uneconomical to operate[2] and was probably too radical in too many ways for the typically conservative corporate aviation market. Some two dozen Starships were produced, and the company had a difficult time selling those. Nevertheless, it was a milestone in digital avionics development in the civil marketplace.

General aviation avionics:
My PC is my copilot

Although the most dramatic advances in avionics computing technologies have occurred in the cockpits of civil transports and military aircraft, the cockpits of general aviation aircraft have also been changing. From the cockpits of globe-girdling airborne corporate suites and VIP aircraft to restored open-cockpit biplanes, pilots are using microchip technology to ease cockpit chores and fly more efficiently.

Large, modern business aircraft such as the Gulfstream G-IVB and G-V, Canadair Challenger and Global Express, and Dassault Falcon 2000 are likely to be as sophisticated in their flight-deck technology as almost any heavy transport described earlier. State-of-the-art display systems and integrated, redundant avionics suites are now the standard. Mid-size corporate jets are now being equipped with flat-panel displays and integrated avionics suites, and many small, cabin-class turboprops and helicopters have a fair share of CRT displays and sophisticated navigation systems. Operators of older business aircraft are finding that to maintain their aircraft's resale values, they must retrofit the cockpit with digital avionics. Indeed, avionics retrofits are among the hottest businesses in corporate aviation today.

Many operators of smaller, piston-powered aircraft are electing to install digital systems made for this category. Although not as robust or capable as their ARINC cousins, these avionics are mimicking many of their talents. Panel-mounted moving map displays, digital readouts, altitude alerters, fuel management system aids, and other aids are quite popular.

Light aircraft are often operated by businesspeople who recognize the utility of personal computers and are comfortable using them. So it isn't surprising that they are taking laptop PCs into the cockpit and connecting them to a GPS antenna. Why? Because a handful of companies have developed some remarkable, all-in-one programs that can turn a PC into a miniature flight department.

PC flight planners, as the systems are called, can be used to create a complete flight plan, automatically gather weather from an online service or the FAA's direct user access terminal (DUAT) system (see Chapter 9), file an FAA flight plan, and, when connected to a suitable GPS receiver, become a moving map display system, showing the aircraft's current location on a bit-mapped image of a NOS sectional/WAC or Jeppesen chart. When the flight is completed, the pilot can enter the pilot and aircraft's flight times into the program's log book utility. A database of airport facilities can be called up to provide a list of fixed-base operators (FBOs) that cater to general aviation operators for fuel, maintenance, car rentals, restaurants, and other services. And when connected to a printer, the system produces a tidy printout of the flight plan, log book page, maintenance record, or just about any other information. Once at the destination, the pilot still has a portable office suite of applications. Such capability was the stuff of fantasy only 20 short years ago.

Data Link: Flying the Information Airway

Data link is one of the hottest topics in aviation today. In the eyes of many in the industry, data link is seen as the cornerstone of much of aviation's future. For others, it's just a new capability to get us there faster and cheaper, and for still others, it might just mean a better selection of movies. But whatever it ultimately becomes, it appears to be a major component of tomorrow's world of aviation computing, so we'll take a quick look at it here in the context of current avionics technology. (Chapter 10 is devoted to the subject of data link.)

For our purposes, data link is any system that provides real-time, or near-real-time communications between aircraft and ATC, aircraft operators, passengers, and the rest of the world. In its most fundamental sense, data link is a basic concept underlying several initiatives to modernize airspace management. The primary goal of these initiatives is the implementation of information systems to improve the safety and efficiency of the world's airspace and related information networks using advanced communications technologies.

FANS

Collins' Pro Line 21, Boeing's AIMS, and other manufacturers' advanced avionics will be the cockpit interfaces to the new communications, navigation and surveillance/air traffic management (CNS/ATM) world that is emerging from various industry initiatives to improve the safety and efficiency of the world's airspace. One of the most significant initiatives is Future Air Navigation System (FANS). This effort is likely to have far-reaching implications for aviation computing systems, especially avionics, air traffic management, and airline operations.

FANS is an airline and ATC industry initiative that began in the mid-1980s to modernize the world's communication, navigation, and surveillance (CNS) infrastructure. Some 60 percent of the world's

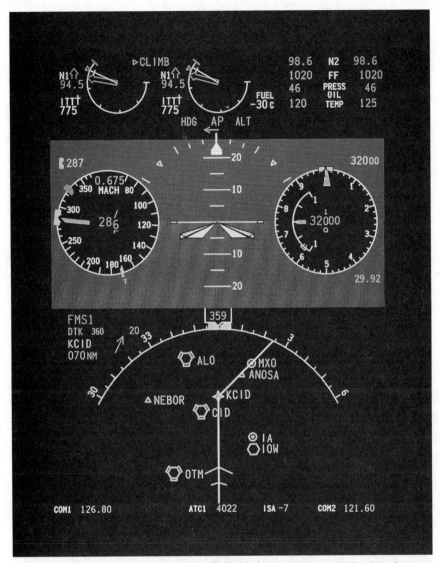

Figure 2.7 A single Collins Pro Line 21 adaptive flight display can be formatted to contain all the information required to fly the aircraft—including altitude, airspeed, attitude, engine data, navigation maps, TCAS, and radar. For fleet commonality, displays can be configured to match existing displays. (Collins)

airspace is beyond the range of air traffic radars, so en route spacing between flights on most transoceanic routes is significantly greater than in areas with radar coverage. This reduces the capacity of some of the world's busiest routes, such as New York/London, Singapore/San Francisco, and Melbourne/Tokyo, for example.

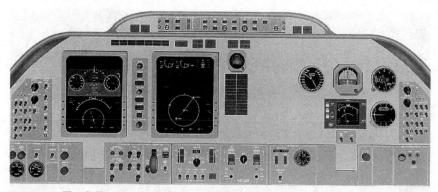

Figure 2.8 The Collins Pro Line 21 avionics suite being used on the Raytheon Premiere I business jet will feature a fail-passive autopilot, a GPS navigation system, and Collins' Integrated Avionics Processing System (IAPS). (Collins)

By integrating the Global Positioning Satellite (GPS) system, enhanced cockpit flight management system (FMS) functions, and satellite communications (satcom) technologies, FANS developers hope to improve spacing and routing of flights and make airspace over remote areas more efficient. The first steps to achieving FANS goals have taken place and real-world implementation efforts are continuing. Again, these are discussed in greater detail in Chapter 10.

When aircraft are equipped with satcom-based FANS capabilities, for example, route clearances can be loaded into an aircraft's FMS via satcom data link, thereby reducing the potential for manual data-entry errors. As the flight proceeds, the aircraft's current position is transmitted automatically via satcom at predetermined intervals and displayed on controllers' traffic management display screens. (Currently, on most long overwater flights, aircraft crews report their positions to ATC manually, using high-frequency (HF) radios, which are often difficult to use due to natural phenomena affecting HF radio frequencies.) Demonstration trials of FANS technology have shown it is possible to reduce separation between flights without compromising safety.

The FAA and several civil aviation authorities (CAAs) in the Pacific Ocean are participating in FANS to bring its advantages online as quickly as possible. The first operational use of direct pilot-to-controller data link communications commenced in June 1995 on a United Air Lines Boeing 747–400 flying between San Francisco and Melbourne.

The transition from the current system of HF voice communications to a system based on satellite data link communication between ATC and pilots is the first step toward implementing FANS. In this application, data link allows for messages, requests, and clearances to be displayed both to pilots and controllers in textual form, reducing the

Figure 2.9 A typical 1960s radio stack, when digital meant "fingers." Despite their shortcomings, manufacturers did manage to produce some very reliable analog avionics.

possibility of errors. The introduction of data link ATC communications will enable operators of data-link-equipped aircraft to take advantage of flexible routings and advanced route planning techniques (called dynamic airborne route planning, or DARP). This in turn will reduce the amount of fuel required to be carried, thus improving passenger and cargo payloads. Additional benefits will include the ability to obtain real-time, inflight weather briefings and communicate with the aircraft's flight/operations department.

The major avionics makers are developing flight-deck systems to take advantage of the expected digital communications interconnectivity that will develop between the cockpit and ATC and airline operations, corporate flight departments, and ultimately, just about any point of connection to telecommunications networks.

Weather on the fly Gaining inflight access to advanced weather products is one of the principal goals of data link developers. The flight crews of airline and corporate aircraft have for years been using the ARINC and AFIS networks (see Chapters 3 and 7 for details on ARINC and AFIS) to obtain character-based weather reports and forecasts while in flight. However, this is seen as an outdated method in today's automated, graphical-computing environment, and it also uses a large segment of the VHF radio spectrum that could be better used for other purposes.

Topping pilots' lists of inflight weather needs is faster dissemination of hazardous weather reports and forecasts as well as the availability of tailored inflight briefings. To that end, the FAA and others have been studying a number of options for delivering weather reports to the cockpit much as it would be received from terrestrial weather networks such as the DUAT system.

One option that was being given consideration several years ago was the concept of piggybacking weather requests and replies on Mode S transponder signals (called "squitters") transmitted from ATC radars. Implementation costs are the biggest obstacle for this concept, however. Further, airline, corporate, and general aviation users say they want to see reduced dependence on air traffic controllers for weather information.

A project studying the feasibility of data-linked weather was conducted at NASA's Langley Research Center in Virginia in 1994, aboard a NASA-owned Boeing 737. In that program, called CWIN (Cockpit Weather Information Network), researchers were able to display ground-provided graphical and alphanumeric weather products, tailored for a dedicated cockpit display unit. Pilots, using a touch-screen interface, were able to call up radar summaries, surface observations, terminal forecasts, ceiling and visibility, and other data, as well as respond to alphanumeric ATC clearances supplemented by synthetic voice messages. Such a system, NASA developers estimate, would cost about $2000 per aircraft.

Other similar programs, conducted in cooperation with avionics manufacturers and university research programs, are demonstrating the feasibility of data-linked cockpit weather display systems. The only thing lacking is a system-wide infrastructure to put it in place, FAA certification standards, and sufficient market demand to spur manufacturers to produce the units.

CNS-12. Another data-link weather system being watched closely by observers of industry technology is CNS-12. Although not quite as sophisticated as the NASA data-link weather program, CNS-12 is about to go into real-world use on some 100 aircraft operated by three USAir Express subsidiaries—Allegheny Airlines, Jetstream International, and Piedmont Airlines.

The carriers are the launch customers for Magellan Systems Corporation's new CNS-12 avionics units. The CNS-12 is a lightweight, integrated ACARS (aircraft communications and reporting system) data link receiver, GPS navigation, and ATC surveillance system. The first installations of the systems were slated for 1996, following final certification approval by the FAA.

The CNS-12 will allow the carrier to provide ACARS communications capability to the USAir Express carriers' Dornier 328 and deHavilland Dash 8 aircraft. Until the appearance of the CNS-12, ACARS capability was typically only found on larger transport aircraft.

Flight crews will be able to downlink out, off, on, and in (OOOI) times to USAir computers, as well as receive predeparture clearances (PDCs), weather, flight planning, and text messages. In the near future, the USAir Express airplanes also will be able to transmit flight-following, performance, and engine trend-monitoring data. The data link capability is provided over ARINC's GLOBALink network and is similar to ARINC service for major air carriers.

The CNS-12 is a growth version of Magellan's earlier CNS-10 unit. The new unit will include a 12-channel GPS receiver that will be upgradable to the FAA's Wide Area Augmentation System (WAAS) differential GPS (DGPS) capability and has the ability to transmit a weight and balance and closeout report via ACARS to a gate terminal. According to Magellan officials, the CNS-12 is the only integrated communications and GPS navigation unit available in the $10,000 price range.

Initially, the unit will only be FAA-certificated for en route navigation and nonprecision approaches, but later upgrades will make the unit compliant with future ATC requirements. An upgrade to a DGPS receiver will add some $5000 to the cost. (DGPS will soon enable users to fly precision instrument approaches at many airports).

The unit comes with a Jeppesen Navdata navigation database and memory for more than 400 flight plans. Flight plans can be downloaded via data card or entered via the unit's keypad. The unit will also provide ARINC 429 interfaces for autopilot, radar, and primary flight instruments. The unit is also being designed to enable corporate operators to install the units across their fleets.

Racal Avionics and Magellan have teamed up to provide satcom capability for users of the CNS-12 when combined with Racal satcom avionics. A first corporate customer of the system is expected to be announced soon, according to Magellan officials.

Inflight entertainment systems

While primary focus of this chapter has been on the concept of the computerized cockpit up to this point, it's important not to overlook what's been going on in the back of the airplane. A whole range of exciting new passenger inflight entertainment (IFE) and communications services, largely based on satcom communications to information and entertainment networks, are being introduced by a number of air carriers and promise to become a regular feature on airlines and long-range corporate jets.

The IFE vendors are having to accept that the airlines are a bit skeptical about many of the vendors' claims, however. In the past, airlines routinely contracted with vendors for IFE systems, only to discover the vendor either couldn't deliver the product on time or was unwilling to support the product adequately once it was installed on customer aircraft. In some cases, the situation was nearing nightmare proportions, as aircraft came rolling of assembly lines with hastily installed IFE systems—or none at all.

To avoid more surprises, the carriers are now imposing industry-wide standards to improve the situation. Airframers are defining a standard interface between IFE equipment and the cabin systems' network on aircraft such as the Boeing 777 and the Airbus 330/340. Standardized interfaces are expected to save the carriers millions by simplifying the task of installing the hardware across their fleets.

The requirement to meet new standards is coming at a time when carriers are realizing their passengers expect them to do more than provide them with a few movies and a seat-back telephone. To meet this demand, the carriers are expected to start buying second-generation IFE systems soon. The carriers also view IFE as a major source of revenue in the near future. Some carriers are expected to retrofit their entire fleets with state-of-the-art IFE systems.

At this point, typical IFE systems in use provide the passenger with an AMLCD screen mounted in the armrest or on a seatback, plus a few basic press-button controls for selecting movies or audio channels from an assortment controlled from a central server. A simple LAN ties the network together.

As the capabilities of satellite communications systems are advanced by satcom communications providers and accepted by the world's air carriers, a host of services are being created to satisfy customer demands that they stay in touch with the world and make productive use of their time while on long flights.

The leading providers of satcom air/ground links are exploring a range of services to meet this need. Passengers will soon find a seat-back terminal can connect them to services that will likely include

- arrival and connecting flight information
- duty-free ordering
- hotel, car, and flight reservations
- pay-per-view video-on-demand (VOD)
- live radio broadcasts
- catalog shopping
- stock market, weather, and sports reports
- flower-ordering services
- electronic games and virtual reality simulations
- e-mail services
- personalized welcomes and alarm calls
- access to the Internet
- inflight gambling

According to one airline industry survey, competition seems to be the strongest driver for the IFE market. Indeed, IFE systems are quickly becoming one of the fastest growing areas of development in airline services. IFE is a burgeoning industry, complete with its own industry conferences, associations, and representation on the engineering staff of major airlines and aircraft manufacturers.

At industry gatherings such as the annual World Airline Entertainment Association (WAEA), the IFE market was showcased with exhibits on the latest advances in inflight telephony, satcom, seat-back display systems, headset technology, full-feature interactive systems, inflight audio, and video programming software. One measure of the success of this market is that CNN's number-one talk-show host, Larry

King, was the guest moderator for a recent panel examining IFE titled "The Seamless Journey: A Look at In-Flight Integration."

The leading satcom providers are exploring how to best provide the services they envision. One philosophy is to offer access to mature services, instead of developing customized services for an unknown market. Such services could include, for example, access to America Online, CompuServe, Microsoft Network, or other online services that provide their own customized services and Internet access.

At this point, however, the size of the market for satellite-based IFE and information network products is unclear. Although many passengers tote their laptops onto airline flights, it isn't clear how many would be willing to plug them into an RJ-11 outlet on the seat back in front of them, nor how many would be willing to spend money to connect to such services. Then, too, what about the passengers who don't bring aboard a laptop or those who are either unfamiliar with or intimidated by navigating the Internet from an airline cabin? Also, there is the cabin crew to consider—how much would such a system interfere with their duties?

One answer might be the use of a central data terminal, but there are logistical drawbacks to that solution. A built-in terminal at each seat might be worthwhile to explore, but such a system would have to be far more sophisticated. Users of today's generation of handset keypads would agree there isn't much appeal to this solution. The passenger data interface will have to provide all the functions of data connections, including credit-card connections and verification, billing, and customer feedback. Another possible interface might be a simple mouse-and-menu system to enable a passenger to navigate through a variety of service offerings.

Ultimately, the IFE and communications market will be better understood, and the hardware and software issues will be settled, since it is obvious there would be *some* sort of market for such services. At that point, it could be said that the passenger cabin was also just another node. Who knows? Maybe the cabin briefing in the future—displayed on the LCD seat-back monitor in front of you—will include the Internet home page address of the flight:

$$http://www.ualflight.601/cabin/@seat32F/html$$

AirShow.   One passenger information product that has long been a fixture aboard many cabin-class business aircraft and in a few long-range air carrier aircraft cabins is AirShow. Among the basic AirShow's many features is a cabin monitor display that is connected to the aircraft's avionics busses, enabling it to display the aircraft's speed, altitude, outside air temperature, and other alphanumeric information. But the most popular AirShow feature is the graphic display of the aircraft's location and route of flight over a colored, moving map. In addition, Air-

Show systems can be configured as an interface to the aircraft's entertainment system (such as an audio, video, or laser disk units).

Now, airline and corporate users of AirShow systems can provide their passengers and crews with a wealth of new information products with AirShow's new Genesys display system. The Genesys service upload, via digital telephone link, provides the latest news and sports headlines, financial reports, NEXRAD weather radar imagery from WSI (a private weather information provider), an atlas and time zone displays, and other features. The Genesys service can be obtained for about $31,500 for a non-AirShow-equipped aircraft or added to existing equipment for $19,500.

Jeppesen and AirShow have recently teamed up to provide integrated information and communications solutions for commercial and corporate aircraft equipped with AirShow. The Jeppesen/AirShow alliance integrates many of Jeppesen's aviation services with AirShow's Genesys data uplink and display capabilities. In addition, the companies will work together to provide enhanced ground-to-air communication and the capability to store and manage such cockpit and cabin crew information as navigation charts, flight plans, weather maps, nongraphic weather information, airport analyses, and other operations and reference documents on board the aircraft, some in CD-ROM format.

Satcom

To connect to the Internet from an airplane cruising across the Pacific Ocean, and since telecommunications are becoming so intertwined with computing applications, you need to become familiar with satellite-based communications systems—satcom.

The first operational satcom telephone-based communications systems appeared in the cabins of long-range air carrier aircraft in 1990. Since then, the market for satcom has blossomed. As of early 1996, satcoms have been installed on some 850 long-range airline and corporate aircraft.

Inmarsat is the London-based international organization that operates satellites providing communications services to aircraft, ships, and land vehicles. Inmarsat was organized in 1982 and now comprises signatories from nearly 80 countries around the world. Each member government nominates an organization within its country (usually a private firm or national telecommunications agency) to be its Inmarsat signatory and service provider.

Major satcom service providers (such as Comsat in the United States, KDD in Japan, British Telecom in the United Kingdom) operate aeronautical satcom ground stations that link satcom users with the "terrestrial" public telephone and data networks. Inmarsat earns its revenue by charging service providers for the use of its satellites. The larger signatories (or groups of smaller ones) are represented on the Inmarsat Council and its advisory bodies and meet several times a year.

Inmarsat Aeronautical (Inmarsat-Aero) is the subset of Inmarsat that provides two-way voice and data communications specifically for satcom-equipped heavy transports and large business jets operating virtually anywhere in the world through the satcom service providers. This service is called Aero-H service. Another Inmarsat-Aero service, Aero-I, is being developed for smaller airline and corporate aircraft. Aero-I is described later in this chapter.

Aero-H. The technical definitions for the types of aeronautical satcom equipment compatible with Inmarsat satellites are defined by ARINC 741 characteristics and the standards and recommended practices being developed by the ICAO. Inmarsat was responsible for defining many of these standards.

The Inmarsat-Aero system comprises three basic elements:

- Inmarsat satellites and their ground support facilities
- Ground earth stations (GES), the facilities that connect airborne users to the international telecommunications networks
- Aircraft earth stations (AES), the equipment onboard the aircraft that is used to transmit, receive, amplify, and encode satcom calls

Inmarsat currently operates four satellites (plus spares) in geosynchronous equatorial orbits at about 22,300 miles (36,000 kilometers) from earth. Geosynchronous equatorial orbit means that the satellites appear to be stationary above a single location above the equator. The network of satellites can "see" virtually the entire globe, except some extreme polar areas. Inmarsat is currently operating its second generation of satellites and recently launched the first of its third-generation satellites. Each generation, as you might expect, has improved capabilities over the previous generation. The Inmarsat-3 satellites will offer "spot-beam" capability, which means they can handle additional calling loads in specific areas of the globe with high volumes of calling traffic.

The satellites provide service over four service regions:

- Atlantic Ocean Region East (AOR-E)
- Atlantic Ocean Region West (AOR-W)
- Pacific Ocean Region (POR)
- Indian Ocean Region (IOR)

Complex ground support systems, based in Inmarsat's London headquarters, are needed to operate and maintain the satellites. These facilities are known as the Satellite Control Center (SCC) and the Network Control Center (NCC). A backup station is operated in Fucino, Italy.

The ground earth stations (GES) are fixed radio stations capable of communicating with aircraft via satellite. These stations are equipped with one or more dish antennas approximately 30 feet (10 meters) across, as well as access, control, and signaling equipment. They provide interconnection between the satellite system and the international telecommunications networks. A satcom-equipped aircraft can thus communicate with any telephone, fax machine, or data terminal anywhere in the world. Communications between the GES and the satellites are conducted in the C-band frequency range of 4 to 6 gigahertz (GHz).

Communications from ground locations, such as air traffic control centers, homes, offices, or vehicles, are provided from existing telecommunications networks to the nearest convenient GES. The GES, in turn, convert the ground-to-air and air-to-ground transmissions into digital code for efficient, error-free transmission to or from a moving aircraft.

Data communications are supported through interfaces with public and private data networks, conform to X.25 and X.75 packet data parameters, and support ISO-8208-compatible data communications. This means that the system can accommodate a PC modem link to a terrestrial computer network or data base.

The aircraft earth stations (AES) are radio sets installed in the aircraft capable of communicating, via satellite, with a GES in the Inmarsat system. This radio link operates on the L-band frequency between 1.5 and 1.6 GHz. The AES can be connected to a variety of onboard systems such as telephones, fax machines, and data terminals.

An AES typically comprises the following units:

- Satellite data unit (SDU)
- Radio frequency unit (RDU)
- High-power amplifier (HPA)
- Low-noise amplifier/diplexer (LNA/DPLX)
- Beam-steering unit (BSU) or antenna control unit (ACU)
- Antenna

Many of these components can be combined into one or two boxes to reduce the space they require. Older satcom systems used to weigh up to about 260 pounds, but newer models weigh around 175 pounds. There are several antenna configurations, which are mounted on the exterior of the aircraft. The most popular ones are mounted on the fuselage or on the top of the vertical stabilizer. There are two types of antenna systems: high-gain and low-gain.

High-gain antennas are used for high-speed data, facsimile, and voice communications. Low-gain antennas are used for low-speed (300 to 1200 baud) data transfers. These are typically used for airline-operational and ATC data link communications.

Voice communications are at the rate of 9.6 kilobits per second (kbps) and generally provide a clear, crisp voice signal. A satcom voice call is usually indistinguishable from a cellular phone call placed on the ground. Most current satcom systems provide up to five voice call channels and one fax line, operating at up to 4800 kbps, although that rate may soon increase. Data rates of up to 10.5 kbps are available. Voice, fax, or data calls made in this way are called circuit mode.

Circuit-mode connections provide an end-to-end communications path, which enables users to apply their own protocols and computer-industry standard Group 3 modems. In packet mode, data is sent in a series of short bursts, or packets. Packet mode is usually associated with airlines using the ACARS (Airborne Communications Addressing and Reporting System) or corporate operators using the AFIS (Airborne Flight Information System).

Satcom applications are likely to expand significantly in the near future. The ICAO has been using the Inmarsat system in trials of various FANS applications to prove the concept of satellite-based transoceanic CNS and ADS. This simply means that satcom can be used to provide satcom-based automated aircraft position reporting and ATC communications in areas beyond the range of standard VHF radio.

Additionally, because of international concerns that the U.S. DOD has ultimate control over the GPS navigation system, the Inmarsat constellation of satellites will be providing a GPS-like navigation infrastructure for Europe.

Aero-I.   Inmarsat's newest service, called Aero-I, was to be launched in the later part of 1996 (to coincide with the launch of the latest generation of Inmarsat satellites) to meet the needs of medium- and short-haul air carriers and light corporate aircraft. Aero-I will offer 4.8 kbps telephony and data service and 2.4 kbps fax capability in selected areas of Inmarsat coverage. Aero-I systems are expected to weigh and cost roughly half that of Aero-H systems. It will be interesting to see how quickly the industry responds to this program and how it evolves.

In November 1995, Inmarsat engineers conducted the first Aero-I voice test call via satellite using recently developed equipment. The call was the first to be conveyed over a reduced bandwidth voice transmission channel, which was designed to maximize a satellite's frequency resources. It was also the first use of new Aero-I coder-decoder (codec) equipment and software. Codec software transforms human voice into a 4.8 kbps digital data stream for transmission purposes and then decodes the message back into human voice at the receiving end.

Inmarsat is adopting Digital Voice Systems Inc.'s (DVSI) advanced multiband excitation (AMBE) voice-encoding algorithm. The DVSI algorithm operates at rate of 4.8 kbps, reducing the bit rate necessary to digitally represent human speech. According to Inmarsat, by using more sophisticated algorithms, it is possible to bring down the bit rate without affecting the quality to the user. The DVSI algorithm offers the best performance for aeronautical purposes, says Inmarsat.

In addition to serving Aero-I, the new voice channel and codec specifications will be compatible with the existing Aero-H avionics standard equipment. The Aero-I test equipment was installed on a Cessna Citation aircraft for flight trials early in 1996.

Aero-I will operate within the coverage of the Inmarsat-3 satellites, the first of which was launched in April 1996. The service will use a new generation of smaller, lighter, and cheaper avionics and antennas. In addition to the standard telephony and fax service for both cockpit and cabin use, Aero-I is expected to support other applications, including news and weather broadcasts, point-to-multipoint data broadcasts, and interactive passenger services.

Data-3. Most ground-side data networks—including the Internet—use an internationally recognized packet-switching standard called X.25. Packet-switching refers to the method of packaging and addressing parcels of data (as opposed to circuit-mode connections) and sending them to their destination, like millions and millions of express-mail packages.

To connect the PC on your lap in the economy-class section of a transatlantic flight, however, will require Data-3, which is Inmarsat's newest method of connecting you to X.25-based networks. Earlier protocols were slow, clumsy, character-based or binary patches. Data-3 has been introduced only in the past year, so many cockpits still use older communications protocols. With Data-3, it will also become commonplace, for example, for flight crews to connect directly into company weather databases and other operational applications, instead of having to request information via the ACARS network (see Chapter 9).

AMSC. Inmarsat might soon have competition in the North American and Central American/Caribbean markets. In April 1995, American Mobile Satellite Corporation (AMSC), based in Reston, Virginia, launched a communications satellite from Cape Canaveral. AMSC plans to offer a variety of satcom-based services, including aeronautical cellular service, through its Skycell telephone gateway receiver facility in Virginia. From there, calls will be directed to existing telecommunication networks for completion. Up-calling to a Skycell-subscribing aircraft is also planned, as are fax and data service and other standard telephone services.

LEO and ICO satellites. In the near future—certainly beginning in 2001—one or more new satellite-based communications networks are likely to be fully operational and will likely be competitive with the Inmarsat system for airborne communications. Some of these networks will rely on constellations of small, powerful satellites placed in low earth orbits (LEO) at 150 to 300 miles up. One plan calls for a constellation of 77 mini-satellites.

Inmarsat, however, isn't going to wait for competition and is developing its own answer to the hand-held satellite cellular gold rush, called the Intermediate Circular Orbit (ICO) network, which will operate at approximately 8000 miles up (12,880 km). This is an important area to watch to see which kinds of services are developed for this medium and which will have an impact on aeronautical information applications.

Through a glass darkly. If the second-generation high-speed civil transport (HSCT) ever rolls out of the factory hangar, it won't likely have any cockpit windows. Developers are confident that they will soon be able to equip the HSCT with synthetic display screens that will replicate, through advanced sensor technology already in use or soon to be, high-fidelity views of the real outside world. Such "synthetic vision" is the alternative to a hydraulic-powered mechanical nose that would otherwise be required so that flight crews could see forward during high angle-of-attack (AOA) flight typically flown shortly after takeoff and on final approach to landing. The droop nose of the Concorde and the Russian TU-144 "Concordski" SST adds a significant amount of weight and mechanical complexity that is only needed for a relatively short period of time, so engineers are searching for an alternative means of providing the pilots with a view ahead.

The synthetic windshield would rely on high-resolution video, infrared and radar sensors, and other sensor technology to allow the HSCT's crew to perceive far more than they could ever see by looking through a windshield. The system would be complemented by side windows.

This chapter has provided only a glimpse at the rapidly changing world of avionics, but I hope that you come away with a better understanding of the new concept of the aircraft as a "node" on the aviation network and a hint at the possibilities this new notion suggests.

References

1. "Dials Versus Tapes," *Business & Commercial Aviation,* March 1996.
2. "In-flight Report: Beech Starship 1," *Business & Commercial Aviation,* September 1990.

3

Wired Wings: Networks, Nav Databases, and the Internet

Long before the Internet became a household word, before the World Wide Web became the hottest thing since tail fins, even before the PC was invented, aviation was already "wired," or networked. In this chapter we'll continue your tour of the aviation computing world by examining the primary telecom networks in aviation. Although they are major elements of the aviation networks, some networks, such as AFTN and the civil and military weather networks, are discussed elsewhere in this book because they encompass other topics. Next, we'll take a look at Jeppesen Sanderson—the major producer of navigation databases used in aviation. The navigation databases company (as well as the similar databases produced by the National Ocean Survey and other governmental agencies) are used almost universally by operators around the world. And lastly, we'll discuss how the Internet is reshaping many aspects of aviation and daily life.

First Nets

Aviation information networks have existed pretty much since the Wright brothers first wired home about their success in the sand dunes of Kitty Hawk. The telephone and telegraph were the medium of exchange for the first intrepid aviators. The first military aviation information exchanges took place during World War I and included information about flight conditions, military intelligence, and combat orders (some of this intelligence was even transmitted via passenger pigeon). Following World War I, the early airlines forwarded weather reports, passenger lists, and cargo manifests to and from their rapidly growing station systems via telegraph, telephone, and primitive low-frequency radio.

The 1930s have often been described as "the golden age of aviation," in that so many pivotal events took place during this decade. The 1930s saw, for example, the first successful—even lavish—transoceanic airline service; the start of our modern ATC system; the first systematic instrument flight; the development of the first jet engines and pressurized cabins; and the beginnings of many of today's global information networks. As we'll see in a moment, the first major milestone in aviation networking was achieved with the formation of ARINC in the early 1930s. Later, SITA emerged as a major feature on the aviation landscape.

Today aviation is one of the world's foremost users of global, networked information services. The primary network users include air traffic organizations (described in Chapter 8, "Air Traffic Management Systems: At a Turning Point?"); the world's meteorological organizations (described in Chapter 9, "The Weather Net and Uncle Sam Online"); military organizations (which use a variety of mostly classified communications networks); and air carriers, which rely primarily on the ARINC and SITA networks for transmitting operational information.

ARINC

The establishment of the airline industry in the 1920s quickly spawned competition for a limited number of radio frequencies. At the suggestion of the Federal Radio Commission in 1929, representatives from several airlines formed a nonprofit company to serve as an unbiased radio-frequency licensing organization. That organization is called Aeronautical Radio, Incorporated—ARINC. Its first headquarters were in the Carleton Hotel in Washington, D.C. The company is now headquartered in Annapolis, Maryland. Then, as now, only airlines could be stockholders, and each could own no more than 20 percent of the shares.

In 1930, when the first four-course radio navigation system and other radios aids were introduced, the Federal Radio Commission gave licensing responsibility to ARINC. By 1935, ARINC had evolved into two independent radiotelephone systems. The first system included some 245 commercial aviation point-to-point radiotelephone links. This network's function was to transmit airline record-keeping and other business information. The second system comprised airport radiotelephone stations, which carried weather information and traffic management instructions. The two systems were made available to general aviation and military users. Airline weather reports were available to the general public.

A few years later, in 1939, testing of air/ground VHF radio links were completed, and the first VHF frequencies were allocated to the airlines.

Figure 3.1 Using a multifunction control display unit, flight crews can receive inflight weather and flight planning information and send operational messages via the AFIS (Aircraft Flight Information System), which is provided by AlliedSignal through ARINC. (Collins)

The first Aircraft Electronic Equipment (AEE) meetings—forerunners of today's Airlines Electronic Engineering Committee (AEEC)—to develop radio specifications were held. ARINC later became a major driving force behind the development of the very high frequency omnirange navigation (VOR) system, as well as the instrument landing system (ILS) used in the United States and many other parts of the world. ARINC has helped to define most of the world's avionics standards.

First links.   The original air/ground links were completed by radio operators who responded to radio on a channel monitored by the operator or responded to data messages or telephone calls from someone on the ground wanting to contact the aircraft.

The radio operators either relayed the message received from the aircraft to its destination or established a telephone connection so that the two parties could speak directly to each other. Radio operators could up-call aircraft in flight only on a prearranged basis. In relaying messages between the radio operator and other points on the ground, it was found very early that typed data messages worked much better than telephone calls, which is why the ground networks of the major aviation telecom providers evolved primarily as data-forwarding, and not telephone, networks.

ACARS.   One of ARINC's greatest technological developments was introduced in 1978: automatic transmission of operational information that was until then transmitted by voice between flight crews and ground personnel. The Aircraft Communications Addressing and Reporting System (ACARS) is an air/ground communication system that enables aircraft to function as mobile data terminals in airline management control systems. This VHF air/ground data link service is used to communicate ATC, airline operational control, and airline administrative control messages between ground-based organizations and the cockpit.

The main objectives of ACARS are improved operational control, economics, and safety. Data can be transmitted automatically, improving the accuracy of reporting times, such as the time the aircraft's doors shut, time away from the gate, time off the ground, time back on the ground, time arriving in the gate (these are often referred to as OOOI times—out, off, on, and in). Additionally, ACARS provides rapid receipt of reports such as fuel status and other management and dispatch information. A small printer in the cockpit enables the crew to send and receive preflight information, such as the aircraft's time setting and weight and balance—all calculated by ground computers, providing procedural simplifications and reduced manual procedures. Some of the safety benefits include timely notification of weather conditions and improved aircraft tracking and flight-following with no additional load on the flight crew.

The ACARS system was an immediate hit with the airlines, saving them a fortune in record-keeping expenses and providing management and dispatchers with a clearer idea of what their aircraft were doing in real time. By 1979, with 134 stations in North America, ACARS provided virtually total coverage in the continental United States, the northern half of Mexico, and parts of southern Canada from 20,000 feet.

With the introduction of ACARS, data traffic could be communicated between air and ground, either point-to-point or point-to-multipoint, without needing a human radio operator as an intermediary. In the 1970s, ARINC and other organizations realized that digital communications were going to be the only way to handle the rapid growth of airline communications, so work began on developing technology to implement a digital-based telecom system. The concept of data link was also studied. In 1970, ARINC installed prototype digital data-link ground facilities in its San Francisco and Honolulu stations. The ARINC centers replaced the teletype printers with visual display terminals and keyboards for data entry. By 1976, all ARINC stations were similarly equipped, and the stage was set for automated data-link capabilities. Technology for a computer-based message-switching system and computer-controlled data link were implemented throughout the ARINC system. These developments saved airline operators tens of millions of dollars in private-line communications charges. As the 1970s ended, ARINC also began work on developing satellite-based data link and introduced ARINC-provided airport telephone services and various other technologies to hold down the airline's communications expenses.

In 1983, a corporate reorganization formed the Arinc Companies, with Arinc Inc. being established as the parent company of Aeronautical Radio, Inc., and ARINC Research Company. In 1985, ARINC completed implementation of an enhanced electronic switching system with the addition of a distributed network, thus creating ESS-DN. DN enabled faster call routings (London to Los Angeles in half a second). This capability ultimately became the ARINC Data Network Service (ADNS), which is a fully digital communications system comprising air/ground and point-to-point message exchanging. The system has become seamless with the recent demonstrations of satellite-based data-link communications on Pacific Ocean routes.[1]

SITA

Paris-based SITA (Societe Internationale de Telecommunications Aeronautique) has been described as "one of the world's best-kept secrets." The low-profile company is by far the largest of the aeronautical communication service providers in the world and has lately begun to reach out beyond the aviation world. ARINC is its counterpart and competitor in the United States and surrounding areas.

SITA's origins are similar to ARINC's. It was started by executives from 11 European airlines in 1949 who recognized a common need for less-costly communications services. The organization's membership now exceeds 500 members and has recently voted to operate more like a true business entity. It formed and has control over an "arms-length" business that provides the actual telecom network services and related marketing activities.

SITA's worldwide backbone network can be thought of as a series of data switching and interface systems (DIS) arranged in a circle, with every circle linked to several other circles. The system, which is also called a "mesh network," uses high-speed DPN-100 switches to meet the airlines' demand for fast, reliable network access. The system is programmed so that if one segment of the network goes down, another can take up the load.

Each DIS is located at or near a large international airport such as London, Paris, Frankfurt, Singapore, Tokyo, Auckland, and so on. The result is a data network that connects all of these top-level nodes together over either direct ("single-hop") or indirect ("multi-hop") paths and provides coverage in 225 countries around the world. Each DIS acts as a hub for several kinds of data traffic collected from and distributed to different kinds of peripheral nodes. These nodes in turn connect to subscriber teletype and computer terminals. The systems are considered open and can be used to transport a wide range of data types. The older character-based teletype communication protocols have been largely supplanted by X.25 and X.28 router-based, packet-switching protocols.

SITA, ARINC, Comsat, British Telecom, and other telecom providers are now in the business of offering all kinds of connectivity services to the airlines of the world, as well as others organizations, such as airline reservations services (e.g., SABRE, British Airways), oil pipeline patrol, and medevac services, as well as terrestrial transportation services. Satcom communications, another aspect of telecom networks, are discussed in detail in Chapter 2.

In addition to its core telecom services, SITA provides airlines, hotels, and automobile rental firms with other support services from SITA's two main data-processing centers. One, a Unisys mainframe-based facility located in Atlanta, provides passenger reservations services for about 100 airlines that cannot or choose not to provide their own passenger reservations services. Atlanta also is home to what SITA claims is the world's largest fare-quotation service for the travel-agency industry and SITA's baggage tracking and handling service, which is used by some 180 airlines worldwide.

As a recent measure of its strength in the telecom marketplace, SITA recently announced that American Airlines has chosen to outsource its entire North American and Caribbean network to SITA. American is only one of a number of carriers that have chosen to outsource their domestic networks to focus on its core business. At SITA's London facility, the company provides airline flight operations support (flight planning, weather, and NOTAMs—Notices to Airmen) to more than 100 airlines on its IBM mainframe computer system.

SITA's answer to ACARS is called Aircom—indeed, to the end user, the services are nearly identical. SITA's Aircom coverage covers the globe with the exceptions of ARINC's North and Central American/Caribbean territory, China, and the CIS (the former Soviet Union), where it has bilateral agreements with telecom providers. SITA is also a player in the satcom marketplace with its membership in Inmarsat, the London-based international mobile satellite organization.

The company has invested heavily in providing new communications protocols and security and has been offering its services to commercial banks, hotel chains, and Internet-based online services such as Compu-Serve. Growth is expected to continue into these types of nonaviation markets. For the first time, SITA has opened itself up to outside investment (the first investor was Morgan-Stanley Worldwide).

The next major telecom development will be to implement a completely transparent aeromobile data network called the Aeronautical Telecommunication Network (ATN). Ultimately, AFTN developers say, it will be possible to send or receive voice, data, fax or communications from anywhere to anywhere.

Airborne command posts

Some steps in the direction of a seamless air/ground network are already being taken, particularly by the manufacturers of corporate aircraft. As the range of some of the latest designs (Gulfstream G-V and Bombardier Global Express) stretch out beyond 6000 miles, it won't be unusual for executives and VIPs to make 12- to 14-hour non-stop flights. This development has raised concerns that the passengers of these aircraft would be out of the communications loop, so the idea of creating airborne command centers has been promoted by the business-jet OEMs to make the cabin a more productive workspace. Executives could be linked, via satcom and the AFTN, to their corporate headquarters, and fully capable workstations could be installed in the cabin, allowing for normal access to corporate databases and communications—even teleconferencing. The aircraft cabin would merely become an extension of the headquarters. At this point, the only hurdles remaining are the currently limited bandwidth of satcom data and fax connections and the willingness of the passengers to routinely travel on such very long-range flights.

Navigation Databases

One of the keys to modern digital flight computing is the ability to store and retrieve electronic navigation data. Until recently, pilots relied on en route navigation charts, bound instrument approach charts, sectional and world aeronautical charts, and other printed navigation materials. Keeping up with the frequent revisions was a major part of a copilot's life, so when electronic versions began to appear in the newest cockpits, it was cause for celebration.

Avionics vendors and aircraft operators of nearly every stripe have significantly increased their use of electronic navigation databases in recent years. All new flight-management systems and GPS navigation units now rely on electronic nav databases. These databases are collections of information that describe, by latitude and longitude, where everything is: airports, navaids, airway intersections, altitude restrictions, special-use airspace, significant terrain, taxiways, runway thresholds, and so on. Electronic navigation databases also provide the digital versions of navigation charts and instrument procedure charts that flight crews refer to when they are flying on instrument flight rules (IFR).

These databases are quite detailed and require frequent updates (just like the paper documents they replaced). Typically, electronic databases are updated every 28 days, although some might require updates only every 56 days.

Other types of electronic databases used in flight operations include databases of hotels and restaurants, fixed-base operators (FBOs), and fuel locations and types. These databases are available from private sources such as Ac-U-Kwik and commercial flight-planning services.

Government nav databases

Government agencies such as the U.S. DOD's Defense Mapping Agency, the National Oceanographic and Atmospheric Administration (NOAA), the FAA's National Flight Data Center, and similar agencies in other countries maintain extensive geophysical databases that are used to produce navigational charts. Government aviation charts are published by the National Ocean Survey (NOS), and they are designed primarily for military and other governmental uses. The NOS has changed its charts to make them more appealing to a broader segment of aviators. The government licenses commercial map and chart organizations to use these databases to produce a variety of navigation products. The largest producer of aviation navigation material is Jeppesen Sanderson.

Jeppesen

Most pilots associate "Jepps" with the loose-leaf binders full of instrument en-route and approach charts that they depend on to visualize
and navigate the world's airways. Pilots of many aircraft that use electronic navigation databases also recognize the name. But there is a
story—and an individual—behind those little black notebooks and
floppy disks.

In 1921, at the age of 14, a boy living in Portland, Oregon, named
Elrey Borge Jeppesen did something that would change aviation history. He took a $5 ride with an itinerant barnstormer and fell in love
with flying. At 20, Jepp earned his pilot's license—signed by none other
than Orville Wright. Jepp bought himself a Curtiss Jenny, which was a
general-purpose, World War I-era biplane, and began his long colorful
career in aviation.

Like many pilots of that period, he began as a barnstormer, flying
from town to town performing aerobatics and wing-walking and giving
rides to anyone who was brave enough to climb into the airplane with
him. He also did some flight instruction and conducted aerial surveys.

In 1930, Jepp signed on with Varney Airlines and later with Boeing
Transport as an air mail pilot. He flew the Salt Lake City–Cheyenne–
Salt Lake City–Oakland route. At $50 a week and 14 cents a mile
(seven cents a mile during the day), this route was the most lucrative—
and dangerous—one the airline flew.

With none of the aeronautical charts we take for granted today, most
pilots used road maps for navigation. When visibility was limited, the
pilots just flew lower and followed railroad tracks, which was called
"hugging the UP," or Union Pacific Railroad. When the weather deteri-
orated too much to continue, pilots often landed on emergency strips—
usually a farmer's pasture. There, they'd wait out weather, alone except
for a sack of mail. Jepp dreamed of one day being able to fly over or
through the weather.

During the winters of 1930 and 1931, Jepp lost many of his fellow
pilots to weather and a lack of published information about the location
of airfields and emergency strips. Wanting to change that problem, he
began making notes in a black, loose-leaf 10-cent notebook. In it, Jepp
recorded field lengths, slopes, drainage patterns, and information on
lights and obstacles. He also included drawings that profiled the terrain
and airport layouts and noted the telephone numbers of local farmers
who could be counted on to provide weather reports. On his days off,
Jepp climbed prominent hills along his route of flight and measured
smokestacks and water towers, noting their elevations in his notebook.

Other pilots heard about "Jepp's little black book" and were constantly asking him for navigational information. The requests became

so frequent that he began to sell copies for $10 each. Pilots were eager to obtain the books and provided him with their own observations on their routes, which he added to the collection of airport and route information.

But reliable charts weren't the only things required to make flying safer. There were no navigational aids then, nor any procedures for flying on instruments. Again, Jepp entered the scene and became involved in the newest technology for navigation—the radio.

Jepp tested new navigational radio aids and developed ways to use the technology for improving point-to-point navigation. He also began designing instrument approach procedures using the information he had gathered on airports throughout the Northwest. These procedures were documented on his instrument approach charts and were the only reliable information in the country for that material.

Jeppesen & Co. began operations in 1934 in Jepp's basement workshop in Salt Lake City. In the late 1930s, Varney Airlines, Boeing Transport, and other companies merged to become United Airlines. United was printing its own charts by that time, but its pilots were buying Jepp's charts. Before long, the airline became the first to subscribe to his service.

During this period, Jepp met his wife, Nadine, who was one of aviation's first flight attendants. Together, they built up Jepp's chart business. They moved to the Denver area in 1941. Over the next few years, Jepp broadened his business to supply flight information publications to the United States Navy, as well as to pilots flying aircraft that provided commercial transportation of people, goods, and mail.

Jepp stopped flying in 1954 at the age of 47 on the recommendation of the company flight surgeon. By then, he was captain of a DC-6 and had logged nearly 10,000 hours in DC-3s alone.

One of his company's most significant contributions to aviation safety occurred in 1947 with the introduction of the first standard instrument approach procedures. Prior to this time, instrument approach procedures in the United States were usually designed by individual operators for their own use and then approved by the FAA, which meant there were as many approaches as operators. Jeppesen and the FAA instituted a program whereby the FAA would prescribe standard approach procedures and authorize operators to use those procedures.

It was a milestone in aviation safety. The first ILS approach chart was published in 1948, and the first VOR approach chart was published in 1949. The first high-altitude en-route charts were published in 1959, which coincided with the first commercial jet service. Area navigation (RNAV) charts were first published in 1971, and profile descent charts were first published in 1976.

Jeppesen was also instrumental in helping to establish the FAA's National Flight Data Center, a central office where data on air traffic control, route structure, airspace, and facilities is collected and disseminated.

As demand grew for charted flight information around the world, Jeppesen opened a subsidiary in Frankfurt, Germany, in 1957. The Frankfurt operation serves the entire eastern hemisphere, including Europe, Africa, Asia, and the Middle East. Working in concert, the Denver, Colorado, and Frankfurt operations provide flight information covering virtually the entire world.

The following year, Jeppesen opened a liaison office in Washington, D.C. Working in conjunction with Denver and Frankfurt, this office maintains the flow of daily data from the FAA as well as plans affecting the future of national and international air transportation.

In 1961, Jeppesen & Co. was sold to the Times Mirror Company and became a part of one of the nation's most successful and highly respected media communications companies. Times Mirror Company purchased Sanderson Film, Inc., of Wichita, Kansas, in 1968. Sanderson Films, under the direction of Paul Sanderson, was a worldwide leader in the production of pilot training systems and aviation educational multimedia materials. Jeppesen & Co. and Sanderson Films officially merged to become Jeppesen Sanderson, Inc., in 1974. The Jeppesen Sanderson corporate headquarters is located in Englewood, Colorado, near Denver. The company recently reopened its office in Los Gatos, California.

Today Jeppesen is recognized as the world's leader in flight information services, flight planning services, and pilot training systems. With operations in the United States and Europe, Jeppesen employs more than 700 people. In 1995, when the new Denver International Airport opened, the main terminal was named after Elrey B. Jeppesen.

Jeppesen is among the world's largest providers of computerized flight planning services, electronic navigation data, and airport analyses. Jeppesen's charts and instrument approach procedures are typically stored as databases on several types of magnetic media (floppy disks and various formats of memory cards and data cartridges), which make them more convenient to use than paper media. However, the little black books are still big sellers. Pilots can now call up en-route charts and approach charts and view them on the primary and multifunction flight displays.

The information that Jeppesen uses to compile its databases and charts is continually gathered from about 185 civil aviation authorities around the world, including the FAA's National Flight Data Center. Jeppesen currently maintains its master database of navigation and related information on an IBM mainframe computer located at its headquarters in Denver. The company's facility in Frankfurt operates as a remote user of the master database. Subsets of data are extracted from the master database for different avionics and system requirements and issued as revisions on a regular basis. Those subsets are then further processed to avionics-specific, manufacturer-proprietary (e.g., Honeywell, Collins, Trimble, etc.) formats that are loadable to an aircraft's avionics systems. Each avionics manufacturer reconfigures the data and specifies the media to make the databases work efficiently in each of their systems.

Jeppesen provides a host of support services for most of the world's airlines, plus more than 100,000 individuals, agencies, and corporate users of its products. Some of these services are specifically marketed to the air carrier and business aviation segments, including

- *Maintenance data services.* Jepp provides aircraft maintenance information via electronic work stations. Revisions are processed as required and are delivered electronically, thus facilitating the timeliness and accuracy of information. Typical applications include aircraft maintenance manuals, illustrated parts catalogs, wiring diagram manuals, minimum equipment lists, structural repair manuals, and fault isolation/reporting manuals.

- *Tailored services.* Jeppesen also creates, maintains, and distributes a customer's data to exact specifications. Typical services include en route charts, standard or tailored, with the user's specific route overlays and instrument approach charts with an operator's operating minima, including ETOPS charts, alternate orientation charts, airport qualifications, engine failure procedures, and so on.

- *Document-management services.* These services include publishing and revising airline policy and procedures manuals, aircraft operating manuals, flight attendants manuals, and similar documents. Many carriers are or soon will be moving to electronic documents that can be accessed via personal computer, computer intranets, and other telecom links.

The Internet

If you're like a lot of people, you've heard a lot about the Internet and all it has to offer anyone with a computer and a modem, but so far, you might not have heard what the Internet has to offer for aviation. The answer is *lots.* But first, let's discuss what the Internet is.

The Internet began in the early 1970s as a Cold War-era electronic communications system to connect a DOD network called ARPAnet, various satellite and radio networks, and computer networks operated by universities involved in defense research. The Internet was an experiment to determine if a network could be constructed that would continue functioning even if parts of it were knocked out in a nuclear attack. You'll often hear the Internet described as "a network of networks," for indeed that's what it is—lots of computers, from university mainframes to corporate minicomputers to commercial Internet servers to individual personal computers.

A primary Internet concept to keep in mind is how things are addressed and delivered. Information is sent in Internet Protocol (IP) "packets" that carry an "address" that any computer can read and forward to any other computer on the network. (Keep in mind that the system was designed to work in the event of catastrophe).

Another important thing to remember: The Net keeps changing as new uses are discovered for it. This dynamic condition is likely to continue for some time and will probably accelerate as new technologies are developed to make use of the Internet. The Internet is a communications medium currently in a stage of development very much like the early days of television, when *The Howdy Doody Show, The Honeymooners,* and *You Bet Your Life* were the big hits. At that time, few would have predicted television would also be used to telecast the first moon landings, the Gulf War, Oprah, or partial nudity on *NYPD Blue.* My view is that the Internet will become a combination of television, a visit to the local library, and the corner ATM. Concepts currently in development, such as Oracle Computer Corp.'s Network Computer (a computer-like device for using the Internet), cable modems, and customized "Web agent" search-engine software are likely to ease the learning curve of navigating the Net and make it even more of a part of our lives.

For now, the Internet can be viewed as a repository of information on tens of thousands of subjects that spans the globe. New forms of information are appearing all the time, and these now include text, databases, software, audio, radio, animation, recorded video, and even live video. It's hard to imagine any other types of media that could be obtained via the Internet, but I won't be foolish enough to say there won't be any new variations. Except for the World Wide Web, which I'll describe in a moment, I won't spend a lot of time describing the other resources on the Net, such as FTP, Gopher, newsgroups, etc. These are worthwhile resources of information—and they are well detailed in the many books devoted to the subject of the Internet.

The Net can be used by anyone equipped with a modem, inexpensive or even free software, and a connection to an Internet gateway, which is called a server. How information is actually exchanged on the Internet is becoming less and less important to the average user because the software required to navigate the Net has been vastly improved over the past few years.

The World Wide Web

The newest development on the Internet is the World Wide Web (WWW). The Web was invented in 1989 by physicists at the European Particle Physics Laboratory. The Web concept is simple. By inserting transparent "hypertext" commands into a document and loading the document onto a server, the author can provide "links" to other pertinent parts of the document—or even other parts of the Internet. When a user clicks a computer mouse on the hypertext word (which is usually set off by different-colored type or underlining), his or her "Web browser" software will "jump," almost instantly, to wherever each of these "hot links" point. One document can link to thousands of others. The software that creates these links is called Hyper Text Markup Language (HTML).

The Web was slow to develop, however, until 1993, when programmers at the University of Illinois National Center for Supercomputing Applications (NCSA) developed a software program called Mosaic, which has since been widely distributed, embellished, and hailed by some Internet observers as the greatest advance in publishing since the printing press. Mosaic and its offspring, such as Netscape, make it possible for anyone to publish HTML-based Internet documents in a variety of attractive formats. These programs can access documents that include artwork, sound and video clips, as well as other Internet resources and utilities.

The latest Internet software is designed to reach out, with a minimum of keystrokes, to all the electronic resources you can imagine—

Figure 3.2 Honeywell's Commercial Aviation Systems' home page provides a link to the division's customer service representatives. (Honeywell)

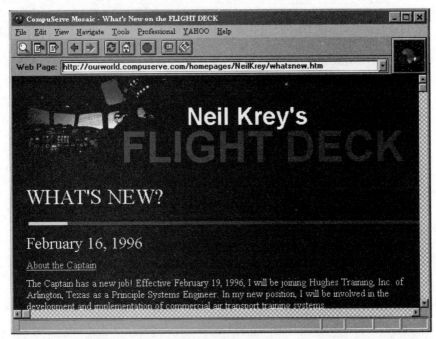

Figure 3.3 Aviation professional are finding all kinds of uses for Internet Web sites. Neil Krey uses his to stay in touch with colleagues in the aviation training community. (Neil Krey)

and more. You can get software for navigating the Internet at retail office supply outlets or download free from the Internet.

Most of these Internet "browsers" come quipped with one or more "search engine" utilities. Search engines are programs that search the contents of Internet directories by keyword (e.g. "radar") or by natural text ("find information on radar"). Some of the more popular ones currently available free are Yahoo, Alta Vista, Magellan, and Exite. Fee-based (per query or by subscription) search engines, such as Infoseek, are also available and offer access to some parts of the Internet that free search engines don't. Most Web browsers also come equipped with customizable, ready-to-launch "hot lists" of popular Web resources so you can quickly locate them again.

In general, Internet costs are shared. The users pay per-hour or monthly access subscription fees to a server that provides access to the Internet. The server, in turn, pays to fund its connections to the rest of the Internet. There is no "Internet, Inc." collecting fees—it's all very democratic.

Although much of the Internet's resources reside on servers operated by academia, the number of private and publicly accessible Internet servers (gateways) is exploding. Internet servers can be found in nearly any city or town in the United States, and many more are opening up all over the planet. Commercial online services, such as CompuServe, America Online, and MSWorld also provide subscription access to the Internet and Web.

Aviation on the Web

So far, we've looked at the Internet in generalities. What about aviation?

In short, if it has, is, or will fly in the future, there's probably information about it on the Internet, plus a whole lot more. There's hardly anything in the aerospace community that hasn't found a home on the Internet.

It is occasionally tempting to complain that the World Wide Web is made up of a lot of "brochure-ware" and vanity home pages with little or no meaningful content, but more often than not, by visiting some of these areas, you'll either learn something useful or gain a new perspective on something familiar. Rather than try to describe all the aviation-related resources (which would be impossible to do in one book and would serve no purpose, since things are changing so quickly anyway), I have, therefore, chosen to highlight just a few of the major categories and describe what you might find under these headings. If you are already familiar with the Internet, you might find areas you haven't visited before, or you might be inspired to set up your own Internet store front.

Aerospace organizations. Many of the better-known aviation and aerospace industry organizations are finding that the Internet is a great way to provide their members with services that either duplicate or complement those provided in other forms. For example, the National Business Aviation Association (NBAA) and the Air Transport Association (ATA), like many special-interest groups, provide public areas in addition to their regular member services that are accessible only by password. The public areas enable the organizations to describe their interests, legislative agenda, policies, and range of services to the public. At most professional aerospace pages, you'll also find publication catalogs, data and statistics, press releases, mission statements, organizational histories, topical descriptions, industry perspectives, and more.

Other types of aerospace organizations include "enthusiast" groups of one type or another—the Experimental Aircraft Association (EAA) and the Online DC-3 Aviation Museum are good examples. These sites might include pointers to aviation museums and collections, air show calendars, historical material, photos, "spotter" forms, weather, usage stats, feedback forms, and sound or video files.

Welcome to the Official Home Page of the
EXPERIMENTAL AIRCRAFT
ASSOCIATION

This site is best viewed with Netscape Navigator 2.0. Download Netscape Now!

The Experimental Aircraft Association (EAA)

Figure 3.4   The Experimental Aviation Association's home page provides visitors with information about the organization, membership, the annual Oshkosh show, a photo contest, and more. (EAA)

Aerospace employment sites. Want to find out who's hiring in the industry or find some new help? Employment sites offer a great way for employers and qualified applicants to locate each other—across town or around the world. As you might expect, individuals with aerospace skills can file resumes in employment databases (usually by annual subscription), and employers can select candidates from the database. This method sure beats filling out application forms. Also, many OEMs have links to information on employment opportunities in their organizations.

Aircraft and equipment manufacturers. Most of the major and some of the lesser well-known aircraft manufacturers have home pages on the World Wide Web, as do a sprinkling of avionics and other aircraft systems vendors. The types of information you'll find include annual reports, division and product descriptions, specifications, software catalogs, press releases, company-related gifts and apparel, surveys, company contacts, employee course catalogs, company histories, company facts, links to co-developers' Web sites, feedback forms, employment opportunities, plant tour information, pilot shops, and links to other divisions.

Aerospace education. A small but growing number of universities with accredited aviation programs, as well as flight and maintenance schools, are making use of the Web to attract new students. You'll find visitors areas, overviews, admissions, course catalogs, departmental and faculty directories, alumni and student organizations, campus maps and photos, financial aid information, virtual tours, aircraft fleet descriptions, calendars, news, athletic programs, research programs, and computer-literacy program descriptions.

Other forms of educational sites include safety-interest pages that focus on one or more topics, such as pilot judgment, aviation medicine, crew resource management (CRM), and other topics. These sites can include searchable databases, BBS listings, pointers to related sites, newsgroups and discussion groups, conferences, and convention information.

Airlines. Hundreds of airlines, ranging from major international carriers to bush-plane operations in remote areas are making heavy use of the Internet. Naturally, the larger carriers are creating splashy, colorful displays with every feature you can imagine. Typical offerings include corporate divisions and descriptions, flight information, schedules, reservations, fare quotes, mileage programs, service information, news releases, employment opportunities, fleet information, corporate histories, seating diagrams, terminal diagrams, cargo services, products and promotions, inflight magazines, inflight entertainment guides, route maps, tour recommendations, management services, feedback forms, destination information, and links to hotel and auto rental reservations systems.

Other airline-related pages focus on various aspects of the industry, such as labor relations, airports, statistics, or stock quotes. One particularly outstanding Web site is Marc-David's Chunk of the Web, which was created by a University of California Berkeley grad student who is studying the airline industry for his dissertation. It is now populated with links to hundreds of carriers' pages and related sites around the world. It also features a utility program that allows you to register to receive a computer-generated e-mail message whenever the page is updated.

Classified advertising. It didn't take long before folks started using the Internet for selling aircraft, engines, parts, training, pilot supplies, accessories, fuel, services, financing, real estate, and just about anything else you can imagine. The better classified sites list items by category and provide forms for posting ads. Some sites are very comprehensive. A/CFlyer, for example, is an online magazine that caters to the world of used business aviation. Besides sales listings, you'll find dealer/broker listings, a product and service directory, industry news, and editors' picks of other worthwhile Web sites.

Aviation Week Group

Figure 3.5 Aviation Week Group (AWG) publishes a number of aviation periodicals targeting various segments of the aviation industry. This shows the recently launched AWGNET site at http://www.awgnet.com. Visitors can read top stories and subscribe to the publications and more.

FAA and other civil aviation agencies. What aviation-related Internet resources are available from the federal government? I don't know yet because I haven't seen them all. Suffice to say there are plenty, and you're sure to find something new every time you visit. Some of the more popular or especially useful sites include:

- *The Official FAA Home Page.* This site is a good place to start tracking down information about FAA programs, facilities, and information. Hotlinks here include news and information, products and programs, the FAA's information gopher, FAA information on the World Wide Web, aviation Internet sites, federal government sites, federal government sites, FAA bulletin board systems (BBS), FAA acquisition reform, and many other links.

- *Aviation Safety Reporting System.* The ASRS provides an overview of this FAA/NASA-sponsored program to collect, analyze, and respond to voluntarily submitted aviation safety incident reports to identify deficiencies in the national airspace system, training, or human factors. The site also provides a program briefing, ASRS reporting forms, information about obtaining records from the ASRS database, the immunity policy, related publications, and links to related sites.

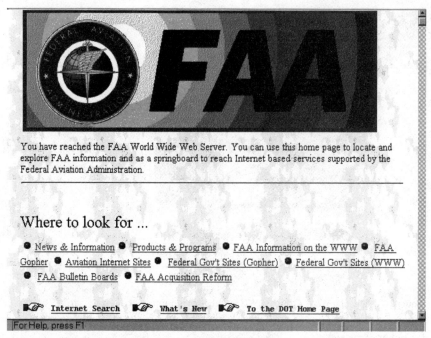

You have reached the FAA World Wide Web Server. You can use this home page to locate and explore FAA information and as a springboard to reach Internet based services supported by the Federal Aviation Administration.

Where to look for ...

● News & Information ● Products & Programs ● FAA Information on the WWW ● FAA Gopher ● Aviation Internet Sites ● Federal Gov't Sites (Gopher) ● Federal Gov't Sites (WWW) ● FAA Bulletin Boards ● FAA Acquisition Reform

☞ Internet Search ☞ What's New ☞ To the DOT Home Page

For Help, press F1

Figure 3.6   The FAA is easy to find on the Web. You'll find many links to other FAA and federal government sites. (FAA)

- *FAA Important Telephone & Fax Numbers.* Need answers to questions on aviation matters? This site is a good place to start.

- *FAA Tech Center.* If you want to learn more about some of the FAA's developmental programs, this site is a good place to visit. Here, you'll find links to general information about current programs at the Tech Center, phone and e-mail numbers, FAA grants and contract solicitations, aviation education, administrator's fact book, employee programs, an FTP server, and more.

Military.   The armed forces are heavy users of the Internet, since they were the among the first to need an Internet. Each branch of the military provides links to numerous public and classified areas. The U.S. Navy's site, for example, provides links to frequently answered questions (FAQs), Blue Angels show schedule, Navy news, public affairs, recruiting, and Internet dial-up services. The U.S. Air Force home page provides links to news, fact sheets, people, pictures, overview, and many other sites—a must stop on a tour of aviation on the Internet. Additionally, other DOD and non-U.S. military organizations have aviation-related pages and other resources.

Internet terminology. The following are some of the more common terms you'll come across on the Internet and online services.

- **Address.** Every Internet location starts with one of the following address codes, which give an indication of what software capabilities are required to access the site's resources:

  http:// indicates a Web site.

  gopher:// indicates a gopher site. (Gopher is the grandfather of the World Wide Web.)

  telnet:// is used for a telnet session. (You log on to a remote computer. DUATS is one example of a telnet session).

  ftp:// indicates an FTP site (file areas).

  news: indicates a newsgroup.

- **Client.** A client is program that requests information from another computer, called a server.

- **Domain name.** The unique, multipart name given to any computer (server) that provides information on the Internet is called a domain name.

- **FAQ.** Frequently asked questions. If you find a resource with "FAQ" in the name, read it first when looking for answers to your questions. Someone has probably asked a similar question.

- **FTP.** File transfer protocol (FTP) is used to locate and download files (documents, sound and video files, software, data, etc.) from Internet sites that provide them to the public. Many FTP sites are poorly organized, so you might want to use another software tool, called "Archie," which can automate the process. The only command you need to know is "find."

- **Gateway.** A gateway is a computer that is used to move data from one network to another. Online services, such as CompuServe, America Online, and Prodigy, are examples of gateways, but there are thousands of others.

- **Gopher.** Gopher is a software tool that makes it possible to browse for resources using menus, then read or access the resource.

- **Mailing list.** A mailing list is a group discussion carried on via e-mail. You "subscribe" to these (usually for no fee).

- **Netiquette.** Netiquette is the etiquette of using the Internet. When you send e-mail, for example, it's considered impolite to use ALL CAPS—it's like shouting.

- **Newsgroup.** Newsgroups are similar in concept to the message boards (discussion groups) and forums that you'll find at the online services. However, since they're scattered all over the Internet, you'll have access to many more topics. However, many newsgroups tend to be less formal—and less organized—than those on the online services.

- **TCP/IP.** Transmission Control Protocol/Internet Protocol are electronic protocols for sending and receiving data over the Internet.

- **URL.** A universal resource locator is the Internet address of a given resource (see **address**). Not all Internet resources are located on the Web; some can be reached more efficiently by entering the resource's URL address into your Internet software. Your software's documentation should provide details.

- **Usenet.** Usenet is a world-wide network of newsgroups that includes thousands of subjects.

- **WWW.** The World Wide Web is the Internet system that enables its users to jump from one Web site to another through hypertext links embedded in Web resources.[2]

The online services

One part of the Internet we haven't discussed much up to this point are the online services. Many people take their first steps in cyberspace on one of the popular online services, such as America Online (AOL) or CompuServe. Each offers a structured environment and thoughtful features that can make one's first experiences in this medium less of a trial-and-error experience than if one were to simply jump out into the Internet with a browser. What's more, the online services have developed relationships with many major news media organizations and other services that make using online services worth the $10 a month they charge for basic service.

Online services also have made it easier to get onto the Net in the first place, by integrating Internet browsers into their access software. Indeed, this feature is causing a whole shift in the online services' focus. For example, where once CompuServe featured a large assortment of electronic message boards devoted to hundreds of topics (e.g., aviation, computing, advertising, journalism, religion, outdoor activities, and pet care) and access to computer manufacturers' help services—and it still does—the company is now focusing its energies on providing Internet access to a variety of commercial Internet "storefront" home pages. This trend is likely to continue. If you haven't yet tried of one of the online services, you might be surprised at what's available for the aviation-minded.

What follows is a guide to what can be found on the online services, but keep in mind that like the rest of the Internet, the content and focus of the online services is changing quickly, so some of this information might be out of date within a year or two, and possibly sooner.

America Online. America Online's aviation lineup includes access to the following areas:

- "Aviation on the Internet" provides access to Internet newsgroups, mailing lists, and World Wide Web "Aero Links."
- Aviation Forum hosts online conferences, such as "Military Training Routes," "Robin' Hood: Simulated Instrument Flight," "Aircraft Minimum Equipment," and "Aviation Trivia Contest."
- "What Is It?" a contest in which you try to identify the aircraft shown in an online image and win a hour of free time online.
- "Aviation Classifieds" are provided as a free service to Aviation Forum members. Categories include a variety of airframe manufacturers, avionics, powerplants and accessories, as well as aviation real estate and employment.
- A monthly Aviation Forum newsletter highlights forum developments, AOL computer bugs, a news item or two, and other information.
- The frequently asked questions file answers common questions (i.e., "How do I learn to become a mechanic?," "The Aviation Safety Reporting System," and "How to use the Aviation Classifieds");
- "Aircraft Gallery" contains a smattering of images of business, commercial, homebuilt/kit, helicopters, and vintage aircraft.
- "Weather News" provides a choice of general-purpose weather images (provided by Weather Services Corporation) that AOL admits won't be of much value for flight-planning purposes.
- "Airshows!", hosted by Frank Kingston Smith, a well-known air show announcer, provides extensive schedules of air shows around North America. The listings are provided by the International Council of Airshows, Inc., based in Jackson, Michigan.

If you haven't already received a copy of America Online's software in the mail, you can find a copy attached to many consumer magazines. Failing that, contact AOL at (800) 827-6364.

CompuServe Information Service. CompuServe's aviation lineup includes access to the following areas:

- AOPA Online provides members of the Aircraft Owners and Pilots Association with news and information about this organization's efforts, as well as the AOPA's Air Safety Foundation, government information, and legislative news.

- Aviation Careers Forum is provided for those seeking aerospace career information. Representatives of a variety of aviation education institutions provide answers to frequently asked questions and discuss topics related to aviation careers.

- AVSIG is CompuServe's oldest and largest aviation forum. It provides topical discussion and online library subsections of aviation topics, including ATC, safety, homebuilts, nonpowered flight, aviation history, corporate aviation, FAA topics, airlines, instrument flying, and others. This forum is managed by the Aviation Safety Institute, which also makes its publications available online.

- AVSUP, a forum also managed by the Aviation Safety Institute, "facilitates professional dialog for aviation vendors and users." The forum's messaging sections can be used to announce or seek information on new products and services, network for career opportunities, as well as discuss general aviation topics.

- Aviation Week Group Online. The online service from *B/CA*'s parent organization within McGraw-Hill provides daily, weekly, and monthly articles published in *Aviation Week, B/CA,* and a dozen weekly and daily aviation and aerospace publications, as well as online conferences and "live news" from major aviation gatherings such as the Paris airshow, the NBAA convention, and other events. This area also provides direct access to A/CFlyer's home page. The related AWG Forum features more than 20 topical discussion and online library sections, including piloting, safety, training, crew resource management, aeromedical, air carriers, business aviation, propulsion, military aviation, career network, legal, aerophotography, and other sections. Much of this material will be moving to its own home page by 1997, but parts of it will remain on CompuServe.

- EMI Aviation Services provides a variety preflight aviation weather and flight planning services for turbine and piston-powered aircraft, including en route weather briefings, fuel burn, and performance calculations based on prestored performance data.

- Flight Simulation Forum is devoted to desktop flight simulators and related gameware.

- Model Aviation Forum, or ModelNet, is devoted to the interests of model builders of all descriptions—airplanes, cars, boats, and rockets.

- NWS Aviation Weather is a direct link to the National Weather Service's aviation weather circuit. It features text-only weather products.

- Ninety-Nines Forum, a private forum, open to the members of the International Women Pilots, is dedicated to furthering the interests of the 99s.

- Women In Aviation Forum was created to help foster greater networking among women with career or hobby interests in aviation and to generate a collective energy among all interested individuals to promote the recruitment of greater numbers of women into aviation and space fields.

- CompuServe also provides direct and indirect links and Net-Launcher software to connect users to Internet FTP, newsgroup, and Web sites.

If you haven't already received a copy of CompuServe's software in the mail, you can find a copy attached to many consumer magazines. Failing that, contact CIS at (800) 848-8990, (614) 529-1340, or +44-(117)976 0680.[3]

Diversities: Other aviation computing topics

With a subject as broad as aviation computing, there are bound to be some topics that don't fall into any neat categories. The following, then, are other subjects related to aviation computing.

Mass storage.   Expect massive increases in computer memory capacity. Terabank Systems of California has led a team of companies to develop an information storage system that reportedly can store the entire contents of the Library of Congress within a tabletop unit.

Gameware.   If you have somehow managed to miss the entire development of computer gameware development since two-player Pong was considered challenging in the 1970s, you're in for a surprise, especially if you've always had a hankering to be a fighter pilot. The latest aviation-based gameware programs feature multiplayer air combat One game, for example, is Kesmai's Air Warrior, which was just being beta-tested on Kesmai's World Wide Web site (*http://www.kesmai.com*) as this book was being written. This advanced multiplayer flight simulator features 27 World War I, World War II, and Korean War aircraft with authentic airplane performance and adjustable levels of play for both veteran and beginner pilots.

Security systems.   Terrorism is all too frequently a part of the aviation landscape. For that reason, antiterrorist authorities are always looking for new ways to deter terrorists. Computer technologies are becoming an increasingly effective tool in this regard, and new ideas are eagerly sought by antiterrorist agencies.

Houston-based Air Security International, a part of the Air Routing International computer flight-planning service bureau, is one company that is focusing on the security concerns of business aviation operators and airlines that operate in the world's more terrorist-prone areas. Air Security provides computerized briefs for operators on airport and security conditions around the world 24 hours a day and provides customized briefs on civil unrest, corruption, terrorism, aviation and flight-safety issues, and healthcare quality, among other topics. Information is updated 24 hours a day, and other services are provided.

Airport and airline officials are taking a close look at Computer Data Systems, Inc.'s Automated Biometric Identification System (ABIDS) as a means of improving security. The Rockville, Maryland-based CDSI says it is the first company to integrate card and biometric technologies into a system that provides positive user identification. The system, now being demonstrated at airports worldwide, would not only "ensure that the right people have access to high-security areas of the airports," but facilitate boarding of ticketless-travel passengers, according to company officials. A prospective ABIDS cardholder would "enroll" traditional identification—such as Social Security number, birth data, address—as required by the customer into a data bank, along with biometric details. These body "signatures" cannot be duplicated and might include retinal scan, fingerprints, and face or hand "geometry"—referring not only to a handprint, but to the shape of the hand, its bone structure, and pressure force. The data then are incorporated into an appropriate card technology, whether smart card (with an information-laden chip), or optical, magnetic, or bar-code cards.[4]

Human factors on CD-ROM. For more than five years, the FAA's Office of Aviation Medicine's Human Factors in Aviation Maintenance research program has published nearly 200 papers, conducted nine workshops, and distributed a variety of software tools for job aiding and training. The agency has released the third in a series of CD-ROMs from the Office of Aviation medicine that, as with previous editions, provides all of the proceedings and phase reports of the meetings on human factors issues in aviation maintenance.

CD-ROM #3 also contains FARs 1 through 200; the FAA Airworthiness Inspector's Handbook (Order 8300.10); and the Air Transportation Operations Inspector's Handbook (Order 8400.10). The disc includes hypermedia software that allows the user to quickly locate any text or image provided in the meeting proceedings and phase reports as well as the FARs and handbooks.

Additionally, there are three new programs on this release: the Electronic Human Factors (E-Guide), the Ergonomics Audit Program (ERNAP), and the CASE Vendor Audit Program. A complete version of the CASE software will be included on CD-ROM #4 of the CD-ROM, which should be available by the time you read this.

CD-ROM #3 also contains two video brochures. One program describes the Office of Aviation Medicine and the Civil Aeromedical Institute. The second program describes the Performance Enhancement System (PENS) field-evaluation plan. PENS is a computer-based tool designed to aid aviation safety inspectors (ASIs) in performing their oversight duties. For the evaluation, PENS will be fielded in all nine regions of the FAA, using four different portable computers (three pen-based systems, one track-ball system). Approximately 36 ASIs will participate in the evaluation, four at each FSDO. Testing the PENS prototype in the field will identify the tools necessary and viable to ASIs and their supervisors.

References

1. ARINC, Inc. 1987. *The ARINC Story.*
2. Gormley, Mal. 1996. "Aviation on the Internet." *Business and Commercial Aviation.* September.
3. Ibid.
4. "Airline Outlook." *Aviation Week & Space Technology.* 1995. September 18:15.

4

Virtual Flight: Simulators and Computer-Based Training

Mention "simulator check ride" to a group of professional pilots and you're likely to instantly raise their blood pressures a notch or two. Never mind the real dangers they might encounter on dark and stormy nights, or the possibility of an inflight emergency occurring sometime in their careers, or even the loss of their medical certificates—a check ride on a state-of-the-art flight simulator is what *really* gets their attention.

In the space of a single flight, a pilot can be faced with a lifetime of hairy situations, including the world's worst flying weather (severe turbulence, icing, and zero visibility); seemingly impossible combinations of mechanical calamities (engine fires during takeoff, runaway autopilots, failed flight directors, improperly entered navigational coordinates and navaid frequencies, etc.); situational ambiguities (autocratic pilots insisting on flying "their way"); inflight medical emergencies (the captain dies during takeoff); and less-than-cooperative air traffic controllers.

In this chapter, we'll take a look at some of the reasons why flight simulators, flight training devices, and, recently, other computer-based training (CBT) systems are considered by many to be the most effective tools for aviation training. Along the way, we'll take a look at some examples of these remarkable technologies to give you a better idea of where they're headed, as well as some of the issues surrounding their use. Lastly, we'll take a quick look at one place where the science of computerized simulation takes center stage.

Virtual Flight Decks

If you've never had the opportunity to see for yourself how true to life a modern flight simulator can be, consider this: The FAA allows the pilots

Figure 4.1 Flight simulators have proven to be the most economical method to train pilots. Advanced Level D sims, such as this one built and operated by FlightSafety International, provide startlingly realistic flight fidelity, including day/dusk/night visuals, audio, full motion, and even artificial smoke to enhance the realism. (FlightSafety)

of certain models of modern transport aircraft to receive their *entire* initial and recurrent flight experience in a simulator. In other words, some pilots can go from simulator to crew member on revenue flights without ever actually receiving any "live" training in the same make and model of aircraft. That's testimony to the creative talents of the engineers and computer programmers who develop flight simulators.

But the primary attractions to using a simulator instead of a real aircraft is cost and flexibility: A Boeing 737 flight simulator costs about $550 an hour to use, versus $3000 an hour to operate the real McCoy, and most maneuvers you can do in a simulator would be unthinkable in an airplane.

Today's flight simulators are a far cry from the Link Trainers that were used during flight training in World War II and for a long time afterward. Indeed, the word *simulator* has different meanings, depending on what you're talking about. An advanced flight simulator is the term used to describe the devices used by airlines and other flight-training providers for initial and recurrent pilot flight training. These devices typically provide highly realistic visual display systems and six degrees of motion (provided by computer-controlled hydraulic "legs" that support the simulator's cockpit mockup).

The FAA defines flight simulators this way:

An airplane simulator is a full-size replica of a specific type of make, model and series airplane cockpit, including the assemblage of equipment and programs necessary to represent the airplane in ground and flight operations, a force cueing system which provides cues at least equivalent to that of a three degree of freedom motion system, and is in compliance with the minimum standards for a Level A simulator specified in AC-120-40.[1]

Simulator manufacturers work closely with airframe OEMs and have developed a number of proprietary software products and computer-based training systems to provide the startlingly realistic audio-visual displays and motion cues. Most simulators are equipped with actual cockpit hardware, including adjustable seats, visors, flight control yokes, throttles, and fully "functional" instrument panels. In addition, most simulators provide realistic audio cues and flight control "feel" that is usually indistinguishable from the real aircraft.

The majority of simulators used today are known as Level "D" simulators in reference to the FAA's definition for a Level D simulator. (That definition goes on for pages, but suffice to say it requires a high degree of realism. A simulator training company representative explained it this way, "In a Level D sim, if you have one wheel go into the grass, you'll feel it in the controls—it's *that* authentic.")

Flight training devices (FTDs) are used by the airlines and other flight training organizations for training purposes that don't require quite as high a level of fidelity as a simulator provides. I discuss this category later in this chapter.

Advanced flight simulators

The advanced simulator market is currently dominated by CAE Electronics, Inc., FlightSafety International (FSI), and Thompson Training and Simulation, and while each vendor has its own specialty, the end result is virtually the same—a cockpit simulator with remarkably realistic effects. Most of the major manufacturers build their simulators to meet the FAA's Level D standard, which provides the highest levels of "fidelity," which is the industry's term for computer-based realism and authenticity to actual aircraft.

To give you a better understanding of how some of the latest advanced flight simulators work, I focus on the simulation technology developed by FlightSafety, but it should be stressed that the other simulator makers' products provide equally capable systems and levels of fidelity. FlightSafety, which has been in the training business since 1951, claims it has the world's largest civilian aviation training organization and operates the world's largest fleet of simulators, with over 160 systems currently operating and more coming off the line. (Canada's CAE has actually sold more systems to end users but doesn't operate as many). FlightSafety's Simulation Systems Division builds the systems for their airline and military customers, as well as its FlightSafety Learning Centers.

The company's accomplishments are significant: FSI was selected to design and manufacture simulation technology for the U.S. Air Force's new four-engine C-17 transport, and FSI has recently designed simulators for the Learjet 31 business jet, Saab 2000 regional transport, and Boeing 777 transport and is developing a simulator for the Gulfstream G-V long-range corporate jet. FlightSafety's list of simulators includes more than 50 of the most popular airline, corporate, military, and general aviation aircraft made in the past 30-plus years.

FlightSafety includes among its air carrier customers All Nippon Airways, Finair, Japan Airlines, Lufthansa, Singapore Airlines, and ATR Training Centers and annually trains some 33,000 pilots and technicians at its network of 36 learning centers in the United States, France, and Canada. Ten of the top 11 major airlines lease blocks of time on FSI simulators, and the company has training contracts with military services and federal agencies such as the FAA, NASA, and the Drug Enforcement Agency (DEA).

Recently, FSI acquired the VITAL Visual Systems computer-generated visual environment technology from McDonnell Douglas and incorporated it into its systems development. FSI has also added aviation technician training, nonaviation simulation training for maritime officers, and nuclear and fossil-fuel power plants operation training to its repertoire. The company's training centers typically operate about 18 hours a day, although some are run around-the-clock due to demand.

Computers in concert. To get a better idea how FSI's simulators work and where the technology might be headed, I contacted John Slish, manager of product information at FSI's Simulation Systems Division (SSD). The Tulsa-based division has been designing and building simulators since 1963.

"Everybody in this business—including us—will tell you they are 'state of the art,'" said Slish. "I will tell you how we do it at Flight-Safety. Let's start first with the host computer. We're using the Harris Night Hawk. That machine comes in several models that we have used on our simulators—the 4400, 4800, and 5800. We typically use the 5800. These are open-architecture, UNIX-based, 32-bit RISC [reduced instruction set computers] computers that are very comparable to an IBM 6000, which is what CAE uses. We actually do software development on an IBM 6000 and just port it over to the Harris."

The simulator cockpit replicates an aircraft cockpit in form as well as function, using a combination of actual aircraft hardware and simulated hardware. The simulator's cockpit shell is a reinforced replica of the shape of the aircraft from the area forward of the rudder pedals to the aft edge of the elevated flight deck. All hardware is mounted into the shell in the same locations as in the aircraft.

The instructor's operating system features three full-color, touch-activated screens for displaying control and navigation data. The system uses an X-windows graphical user interface (GUI) system. Virtually all commands are accomplished with the touch-screen menu of commands, and only a minimum of system knowledge is required to operate the simulator's systems. Control of the simulator is accomplished from a single terminal. The hydraulic power unit (HPU) provides filtered, temperature-controlled hydraulic fluid to operate the simulator's control loading and motion systems.

For most applications, SSD uses FORTRAN 77 for the simulator's math models, but if they are developing something for the military, the programmers use Ada (see Chapter 2). The computers can handle either language, and the simulator's software is written in C.

FSI flight simulators are comprised of a collection of subsystems that work in concert, under the direction of the host system. Other subsystems include

- The host computer
- Control loading/motion

Figure 4.2 A view from the right seat of an advanced aircraft simulator. The world's flight simulator fleet is expected to increase some 70 percent in the next decade. The largest growth will be in the Asia-Pacific market. (FlightSafety)

- Instructor station (X terminal)
- I/O file server
- Avionics interface
- Weather radar (WXR)
- Cockpit I/O
- Visual subsystem

SSD uses Motorola processors (typically a 68040) in most subsystem processors. Some subsystems, such as the control loading system, require up to eight of these processors.

The host computer issues calls to each of these subsystems to produce the desired effect in unison with the output of each of the other processors. This entire process—called the refresh rate—takes place 60 times per second, or 60 Hz, or less, depending on the complexity of the events taking place, and indeed, many of the FSI simulators run at about 30 Hz. (The individual subsystems, however, process at a much higher rate—up to 5000 Hz is not uncommon for the control loading subsystem.) The real-time communications for this system takes place between the host computer's bit 3 board and the bit 3 board in each subsystem.

One of the most important components of the simulator is the file server. It provides a single repository for all the simulation software, reducing the peripheral requirements on the other computers in the simulator complex and allowing for a single point of configuration control of the software. It also simplifies the task of backing up the system's software. When the simulator is initially turned on for the day, each subsystem, in turn, goes to the file server and takes out a copy of the software required to run that subsystem. This process takes place on the system's Ethernet. Access to the subsystems by FSI personnel is accomplished with X-window terminals.

Holy smoke! At the heart of the system is the digital control loading system (DCLS), which employs FlightSafety's digital remote interface to accurately simulate control forces and travels. The digital motion system (DMS) activates the simulator's hydrostatic actuators—the large hydraulic legs that provide the actual sensations of motion to anyone in the simulator. The motions are computer-controlled to keep the motions in sync with inputs from the flight crew.

The avionics interface computer (AIC) interfaces standard aircraft avionics to

- The host computer.
- The electronic voice processor (EVP) system. The EVP is used to simulate aircraft radio reception. Variable rate, generic, or airport-specific background chatter for tower, ground, approach, departure,

and center frequencies, as well as dynamic ATIS (Airport Terminal Information System) chatter, are provided through communications channels. To further enhance realism, noise and signal strength are controllable from the host computer. The EVP collects data from the host computer on the current weather, airport in use, and so on, and then formulates an ATIS message in real time. If the formulating data changes from the host, then a new ATIS message will be issued on the fly and in sync with current simulator flight conditions. The speech used for the EVP is digitized from actual sources and is stored on a hard disk for playback.

- The environmental sound system produces ambient sounds produced by the aircraft, such as wind, landing gear and flap extension/retraction, and engine noise.

- The weather radar simulator provides the appropriate flight displays with simulated weather radar indications. The weather radar simulator is equipped with a library of 10 generic storms that vary in intensity from areas of light rainfall to intense thunderstorms. Up to five storms may be active at the same time, and these can be combined to form squall lines. Storms can be given speed and direction.

The use of the AIC allows for unmodified aircraft avionics equipment to be installed on the simulator and translates data produced by the host computer to formats required by the avionics. On FSI's Boeing 777-200B simulator, two AICs are used, one dedicated to ARINC 429 and circuit breaker activation and one dedicated to ARINC 629.

Another key component to the simulator is the visual system. The simulator industry uses a number of methods to generate the incredibly detailed visual displays. Until the early 1990s, simulation visual displays were limited to night-time displays (which are quite impressive to begin with), but the software engineers of today's latest systems use a host of technical wizardry combined with real photography to create feature-rich cockpit views. The visual displays used are usually customer-specified. All Nippon Airways, for example, uses displays of Tokyo, Singapore, and San Francisco, its primary destinations. However, each simulator is equipped with a global database for all runways greater than 4000 feet.

Additionally, many of the FSI simulators are equipped with an instructor-controllable smoke generator. This nice little touch releases a white, nonpungent smoke into the flight compartment for training in emergency conditions. The smoke is harmless to the simulator occupants and doesn't leave any residue. An exhaust fan is able to remove the smoke after an exercise, and the simulator's smoke detectors are overridden during the exercise.

Flight crews and simulator instructors are given headsets, boom mikes, and hand mikes, and the instructor can speak to individual pilots, play the role of ATC, or select audio channels to play simulated ATIS messages or passenger cabin PA announcements.

The Night Hawk 5800 runs at 50 MHz, can support up to 128 Mbytes of memory on each processor board, and can be configured with up to 512 Mbytes of global memory. The computer uses the Harris CX/UX operating system. CX/UX is a low-overhead, high-performance version of UNIX specifically designed to provide the performance required for real-time applications, such as a flight simulator. A digital audio tape unit is used for backup of the mass storage disks. The system operates on an Ethernet local area network.

Future sim. Full-motion, audio-enhanced, day/dusk/night visual continuity—some would even describe the experience as a form of virtual reality. It might seem like today's flight simulators are reaching the pinnacle of fidelity.

Not quite. Experts in the simulation field predict continuing advances in such areas as aircraft performance matching, even greater visual realism, crew training scenarios, and instructor capabilities. FlightSafety's Boeing 777 and Gulfstream G-V Level D simulators, currently being developed by FSI, Boeing, and Gulfstream engineers, provide a preview of things to come.

They share the same major technical enhancements, including an advanced instructor workstation and FSI's new-generation VITAL 8 ChromaView visual system—the system now in use in All Nippon Airways' Boeing 777 simulator. Indeed, that is the first installation of ChromaView technology, which is a significant advance over FSI's previous VITAL 7 visual display. The VITAL 8 system has subsequently been used in flight simulators developed for business jet simulators, such as the Gulfstream G-III, G-IVB, and G-V, Dassault Falcon 50, 900, and 2000, and the Learjet 60, as well as the Bell 412/212 and 430 helicopters.

The ChromaView system improves the visual capabilities of a simulator by providing a panoramic scene that measures up to 225 degrees horizontally and 40 degrees vertically, as well as more realistic weather conditions. Actual aerial photographs of airports are digitized with enhanced images of prominent landmarks to create a startling degree of visual realism. This realism also extends to ground operations as the crew taxis to and from the active runway. Ground vehicles move about and ramp personnel guide the plane into the jetway. All Nippon is apparently pleased with the VITAL system—it recently ordered one for its Boeing 737 simulator.

Other simulator improvements to look for in the future include still-better control feel and performance simulation. Some of these improve-

Figure 4.3   FlightSafety International's new ChromaView visual system is being used for business-jet and air-carrier flight simulators, such as the Gulfstream G-V and the Boeing 777. (FlightSafety)

ments will come with improved flight performance data from the manufacturers. Software programs derived from this information are the key to matching the aircraft and simulator performance.

Great strides are now being made in the instructors' workstations, too. New software is in development that will greatly expand the range of training situations that the instructor can call up. Either manual or programmed events can be selected to simulate repetitive instrument approaches or LOFT (line-oriented flight training—a euphemism for on-the-job training with preplanned objectives). The good news is that the instructor will be able to spring up to 600 malfunctions on training crews. The bad news is that the instructor will be able to spring up to 600 malfunctions on training crews!

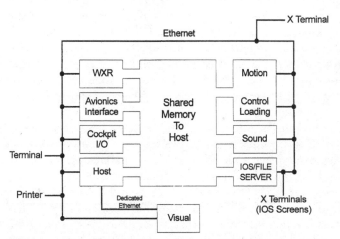

Figure 4.4   Functional block diagram of a FlightSafety advanced flight simulator. (FlightSafety)

FTDs

While simulators are remarkably effective at training flight crews, they're very expensive and can be tough to maintain. Flight training devices (FTDs) offer a less-complex and usually less-costly alternative. In many cases, FTDs are the only choice. Most FTDs depend, to a large extent, on computer technology.

The FAA defines FTDs this way:

> An airplane flight training device is a full scale replica of an airplane's instruments, equipment, panels and controls in an open flight deck area or an enclosed airplane cockpit, including the assemblage of equipment and programs necessary to represent the airplane in ground and flight conditions to the extent of the systems installed in the device; does not require a force (motion) cueing or visual system; is found to meet the criteria outlined in this Advisory Circular for a specific flight training device level; and in which any flight training event or flight checking event is accomplished.[2]

The FAA further defines FTDs to delineate the varying levels of fidelity to real aircraft cockpits and subsystems. The most advanced FTDs closely resemble cockpits of actual aircraft, while others can be generic to a certain category. Most FTDs have no motion systems, but some are starting to use very sophisticated visual displays that rival the visual fidelity found in some advanced, full-motion simulators. Ultimately, however, final approval of a specific simulator or FTD is at the discretion of the FAA principal operations inspector (POI). The POIs tend to refer most of these approvals to the FAA's regional office in Atlanta, Georgia, which is responsible for approving flight simulators.

Simcom.   Today, the price of some models of previously owned, cabin-class recip twins can be a fraction of the price of a new light twin. And a used turboprop twin can be had for only a little more than that. Aviation insurance companies generally require pilots of these big twins and turboprops to undergo formal transitional, initial, and recurrent training, either in an aircraft or in a simulator, depending on the insurer's assessment of the operator's risk. In general, insurers prefer the use of simulator training.

The principal companies offering simulator training for pilots of these aircraft are FlightSafety and SimuFlite. These companies base most of their training programs on classroom instruction and full-motion FAA Level C flight simulators. (Although significant in a technical sense, the differences between Level C sims and the Level D sims discussed above are subtle and insignificant to most pilots who use them.) Full motion is expensive to buy, certify, and maintain, so motion-based, model-specific simulator training in many light piston and some turboprop aircraft simply has not been available elsewhere. Ground-based training for such airplanes has often been accomplished in what the FAA defines as flight-training devices.

The arrival of low-cost, computerized flight simulation software and new forms of visual cueing systems has led to the development of new products that can provide pilots on a budget or other needs with training alternatives to full-motion simulators and the programs offered by some airframe OEMs and the leading "sim schools." General aviation training device and simulator manufacturing companies have developed very capable and effective nonmotion instrument and cockpit procedures simulators for single- and multiengine piston, turboprop, and a few jet aircraft. Many of these devices can be found in hangars and classrooms around the world.

One example of a company that has created a collection of impressive computer-aided FTDs is Simcom, located in Orlando, Florida. Simcom offers training for a number of light twins, including the Beech Baron series, a number of Beech King Air models, a number of Cessna 300- and 400-series twins, Piper Navajos, Cheyennes, and Aerostars, among others.

Each Simcom FTD started life as a real airplane that ended up in a salvage yard. Simcom acquires the front end of the fuselage, including the cockpit area. The hulk is then "crafted" into a FTD, complete with functional flight controls and instrument panels. The control yokes and rudder pedals are connected to servomotors, and new wiring and instruments are installed in the cockpit.

Next, the cabin is completely refurbished, and the entire assembly is mounted on a fixed platform that houses ventilation and the FTD's computer system. The unit is then mated to Simcom's proprietary visual display system. This system rivals those of Level D simulators. It features a 170-degree-by-40-degree view from the cockpit. With a transport delay of 90 milliseconds, it can replicate day or night in various weather conditions. These capabilities compare favorably with the FAA's requirements for Level C and D simulators, which require a 150-degree-by-30-degree view and a 150-millisecond transport delay. Simcom's FTD presents the student pilot with simple, fairly realistic, computer-controlled polygonic landscapes projected on large screens outside the cockpit windows with ample scene content and detail to be realistic—with no significant parallax distortion.

Video software matches the windshield view to instrument indications, cabin sound, and flight-control responses. Indeed, I found the control feel on Simcom's King Air B-200 simulator strikingly authentic in all flight regimes except the landing touchdown, which was more of a thump—and nearly identical to the touchdown of a FlightSafety King Air simulator I had flown previously. As in most simulators, the instructor plays the role of an air traffic controller and can "fail" any of

Figure 4.5 Instructor workstation on a Simcom advanced flight training device. (Simcom)

the systems from a workstation in the back of the simulator. The instructor monitors the students' flights on a CRT display, which can be played back for review.[3]

Xionix FMST. Another benchmark example of a FTD is the Flight Management Systems Trainer (FMST) produced by Xionix Simulation, Inc.

Figure 4.6 The view through the cockpit of a Simcom Cessna 400-series flight training device. Although this system lacks the motion base of a Level D flight simulator, the sights, sounds, and control yoke sensations are sufficiently authentic. (Simcom)

The FMST has attracted the attention of a number of major air carriers, including American, Cathay Pacific, Continental, United, All Nippon, and many other operators of Boeing 757/767s. FMST allows these operators to train crews in the operation and interaction of the flight management computer (FMC), electronic flight instrument system (EFIS), and autopilot systems found on these aircraft. There are numerous variations on the trainers and the program's architecture is designed to be very flexible, allowing for customization for each training establishment.

Through an agreement with Honeywell, FMST uses actual Boeing 757/767 Honeywell flight management computer software and represents the functionality of the aircraft FMC. Using the aircraft FMC operating program and navigation databases makes updating the trainers easier and ensures agreement between the trainers and aircraft. Three system architectures are available with FMST. The simplest version performs the functions of an FMS; the second, called the FMCEmulator, uses a Honeywell database that is used on the customer's aircraft; the third uses the aircraft's actual flight management computer and associated avionics.

FMST also comes in three configurations. One uses a Silicon Graphics' UNIX-based "Iris Indigo" multimedia workstation. The trainer includes a simulated FMS control display unit and mode control panel, which interface directly with the workstation. All other components are displayed graphically on the workstation's CRT display. A second configuration adds a single set of EFIS displays (EADI/EHSI) and is a self-contained unit that can be rolled around on casters. The third option is a full-scale mock-up of the airplane's forward instrument overhead panel and center pedestal. Both the captain and first officer positions are equipped with active EFIS displays, control display units (CDUs), and FMCs. Nonfunctional system control panels and indicators are displayed as full-scale photos or 3-D mock-ups. The system's cockpit displays include many "real-life" features, such as a functional airspeed bug (a reference cursor to "set" a desired airspeed on the EFIS), a digital readout of the Mach number/airspeed, and a digital readout of altitude. Numerous options are also available.

Free play is among the FMST's strongest features—users can take the system through all phases of a flight plan entered by the user, including ground setup, takeoff, climb, cruise, descent, arrival, approach, landing, and go-around. The system comes with the performance characteristics of the airplane that interact with the FMC to provide correct estimation logic, climb/descent profiles, cruise, holding speeds, takeoff and approach speeds, and other maneuvers. All active components are realistic, simulated replicas of actual aircraft units. Switch travel and tactile feel are virtually identical to the aircraft. The program also provides rapid positioning and repositioning modes, as well as snapshot and compression modes for postflight analysis.

Virtual walkaround. A number of the well-established flight-training organizations have recently begun to offer new approaches that use the power of the computer to make learning more compelling while holding down costs. The new medium replaces the traditional blackboard, filmstrip, and flip-chart presentations of the past with computer-assisted multimedia learning centers. These systems replace traditional classroom audiovisual displays with large-screen projections of computer-generated images combining full-motion video, photographs, animated schematics, system block diagrams, illustrations, and other materials in one easy-to-use classroom presentation system.

The new systems enable classroom instructors to present real-time cockpit procedures, filmed in an actual cockpit, with detailed animations of the systems being discussed. The new tools are particularly effective for teaching emergency procedures, where errors can quickly be compounded.

Advanced computing systems are used in other ways by providers of simulator training. At FSI Training Centers, for example, clients can now receive some of their training with the company's Animated Classroom Presentation System (ACPS), which has been introduced at many FSI training facilities.

ACPS has been described as the ground-school equivalent of simulator flight training. The Proxima 8300 LCD projector affords a large display that can include computer-generated aircraft preflight and postflight walkaround inspections and system schematics that show flow patterns. By integrating cockpit switches and a clear visual display of how flows are altered, ACPS makes clear the workings of even the most complex systems.

Other classroom advanced-technology developments include programs such as FSI's multimedia IDEAL (Interactive Databased Environment for Accelerated Learning), which combines multimedia computer displays and databases to present material. Similarly, a multimedia learning system developed by SimuFlite Training International, a Dallas-based simulator training provider, replaces traditional classroom audiovisuals with a large-screen projection of computer-generated multimedia images combining animated schematics, detailed illustrations, and video. This system gives pilots-in-training a balance between procedures and systems.

CBT

Recent advances in computerized training systems are causing a lot of people to believe the personal computer will be the next major platform for aviation training. In fact, it already is. The computer-based training (CBT) category includes programs that can be used to sharpen and maintain one's instrument flying skills (such systems are becoming known as PC-FTDs) or to enhance or even replace traditional classroom training on aircraft systems, avionics, basic airmanship, or a host of other topics that lend themselves to graphical computer treatments.

As you've seen, an advanced flight simulator can cost millions of dollars to purchase and maintain, and regular six-month visits to a training organization is no small chunk of change for an operator on a budget. But the cost—not to mention the risk—of training in real aircraft makes that option even less appealing. However, new alternatives are emerging that are bridging the gap between costly simulators and FTDs and traditional classroom and read-the-flight-manual training. In one area—advanced avionics—CBT systems are gaining favor with operators who can't afford to park an airplane while new crews practice using the airplane's FMS.

Although few developers of CBT systems have thus far convinced the FAA that using PC-based CBT systems is an effective and economical method to provide flight crews with recurrent training, other products are emerging to meet the demand for economical training support alternatives.

Computer-based training aids first appeared in the late 1970s, but they were usually costly, proprietary minicomputer systems that amounted to little more than "page turner" programs—click on a subject and an image of a landing gear, for example, would appear in a simple three-color display. Click again and the gear would retract. An auxiliary display might show a barber-pole graphic to indicate the hydraulic fluid was pumping. Due to the snail's pace of early minicomputer processors, the whole procedure sometimes took longer than the real event—and was about half as interesting. A multiple-choice quiz followed. Snore-ware, to be sure, but it usually beat traditional classroom methods or plywood-cutout cockpit mock-ups. (*Flight training devices* wasn't a term anybody used seriously in the mid-1980s).

With the arrival of graphical PC-based computing in the late 1980s and early 1990s, a handful of enterprising training system developers discovered they could quickly and inexpensively produce training software programs. At about the same time, computer gameware developers (with Microsoft at the head of the list) were having phenomenal success in selling programs—to nearly anyone equipped with a PC and a joy stick—that could approximate the cockpits of simple training aircraft such as a Cessna 150, or, for the more adventurous, the cockpits of a fighter, a Boeing 747, Baron von Richthoffen's famous WWI tri-plane, or even a space shuttle. The market was ready for serious training tools that perform with as much realism.

PC-FTDs

Pilots of light, single- and twin-engine recips often don't have the luxury of advanced flight management systems, integrated flight controls, or even a copilot to share the workload when flying in weather. At the same time, most pilots of these aircraft rarely use their instrument flying skills. More often than not, when the weather's lousy, they'll find some other means of travel or won't go at all.

It can be argued that these pilots are in more need of sharp instrument flying skills than most other pilots. Yet to stay sharp usually requires going up in a real airplane and shooting approaches and practicing other maneuvers. That can be expensive and sometimes downright dangerous for a lot of reasons. The only other FAA-sanctioned option is to use one of the few FTDs designed for this category of aircraft. These FTDs have seen improvements over the years, but they are not always convenient to use, and many pilots don't find them suitably challenging after a few uses. They often are viewed as "old tech" when compared to today's "desktop flight simulators."

So it isn't surprising that many pilots have bought or rented time on a desktop PC instrument procedures program. This category includes programs that emulate the look of generic or aircraft-specific instrument panels and, to some degree, the performance characteristics of airplanes. The airplanes range from light singles to heavy transports.

The makers of the software provide a whole range of peripheral details, including extensive navigation databases, computer gameware joysticks in the guise of flight controls—rudder pedals and all—as well as plotters for tracking and evaluating one's progress after the flight, and even scenery one might see through the windshield. I've been told that some users of these programs connect, via modem, to users of ATC "simulation" gameware programs to enhance the learning experience. Other gameware users join up with those with similar interests in simulated dogfights. Some PC-FTD makers have provided multiple CRT displays of various avionics panels in an effort to provide greater realism to their products.

The instrument displays provided on most of these programs are as responsive (maybe even more so) as a real airplane to control inputs, and it is easy to understand why so many people use them. Indeed, entire Internet and online service discussion areas have grown up around the topic, and heated arguments (usually between the vendors) often arise over which of the many programs provides the most authentic flight experience.

The users and makers of the "serious" training-oriented programs (as opposed to gameware programs that sometime resemble electronic arcade games) recognize how valuable these tools can be, but the FAA hasn't budged: To date, the agency won't allow pilots to credit any of the time they spend using these desktop sims toward meeting the legal requirements for initial or recurrent instrument experience. This stance flies in the face of several studies of the effectiveness of PC-based instrument proficiency training systems that have shown them to be valuable learning tools.

Happily, there are signs that the FAA is beginning to explore the issues raised by the use of PC-based FTDs. The first FAA-sponsored conference on the subject was held in October 1995, so it would appear the agency is struggling to define what standards might be needed to qualify approval of at least some desktop sim experience. Several industry groups are also working on the subject, but as yet a consensus has not been reached. Nevertheless, it's an important area to watch.

Finally, what business jet captain, weekend GA pilot, or pilot wannabe hasn't daydreamed of piloting an F-16 Eagle, a Spitfire, or a space shuttle? The immense popularity of PC-based "flight simulators" has generated magazines, online service discussion groups, Internet

Web sites, and conventions. These flight sims are fun, challenging, inexpensive, and can be run on most PCs.

CTS. One of the first companies to develop PC-based systems training was Computer Training Systems (CTS) of Tenafly, New Jersey. Since the mid-1980s, CTS has been producing interactive, PC-based animated graphical training programs on aircraft systems (such as electrical, hydraulic, fuel, engines, and caution panels) and aircraft flight manuals for more than 200 aircraft. The programs feature integrated testing and record-keeping modules and pilot indoctrination materials. CTS currently produces in-depth, model-specific recurrent training "system simulators" for more than 200 aircraft, as well as CBT software for GA subjects. Individual modules provide graphic displays of aircraft systems and the automatic record-keeping program monitors students' test scores.

CTS was recently awarded a contract to supply computer-based training to more than 100 FAA Flight Standards District Offices (FSDOs). The contract specifies CTS to provide "a large number" of aircraft system simulators and more than 100 copies of its general subjects (indoctrination) programs. The CTS general subjects program includes material on the FAR/AIM, anti-ice and deicing systems, aviation physiology, Jeppesen and NOS charts, windshear, weather, survival, first aid, hazmat handling, and other subjects.

The company's software design isn't as consistent as it could be and could use some updating. To its credit, two-layer password protection is provided, and the testing and record-keeping module is better than I've seen on other CBT products. This feature alone would be appreciated by administrators of large flight departments that do a lot of initial and recurrent training and need in-depth training documentation. Program documentation consists of administrator's and student's instruction manuals.[4]

Ingenuity and opportunity

Other companies and individuals also have developed personal computer-based training programs for one or two specific models of business jets, so I suspect their creators are corporate pilots who might have long layovers and an interest in computing. That is not to say these programs aren't worthwhile—many of these systems show a lot of ingenuity and interesting features. Some of the programs are lavishly illustrated with animated and well-labeled color photos (actually, bit-mapped images), illustrations, and photos-within-illustrations that can be activated with the click of a mouse, as well as interactive text dialog boxes, pop quizzes, and flash cards. Some programs furnish the user with a final exam that keeps score as you go, and one vendor, for a small fee, will create customized tests for the user. The text that accompanies the graphical content in some of these programs is almost compelling, which says a lot for a subject that could be deathly dull the second time through.

One program (which I won't mention by name because the maker is embroiled in a legal hassle with an avionics maker over copyright and distribution issues) provides a great way to learn how to use a popular high-end flight management system found on corporate jets. When this training program is launched, a navigation program runs continually in the background, moving through space in relation to the real world to simulate flight. Another module simulates the aircraft in motion, providing aircraft "hardware" information (heading, altitude, airspeed, and power settings) to the program's FMS emulator and the program's simulated air data computer. The software also receives data from an onscreen, simulated control display unit (CDU) authentically resembling that of the FMS. The software includes a tutorial module that looks like a spiral-bound notebook next to the CDU. You can build, "fly," and modify flight plans, including VNAV waypoint intercepts and holding patterns on the fly, and observe your progress on an EHSI.

NASA's laptop pilot training aid.   A NASA Ames research scientist, Steve Casner, has developed a prototype software program that helps pilots learn how to fly advanced commercial aircraft using a laptop computer. The program, operating on Macintosh computers, mimics the FMS found in large transport aircraft and enables flight-training professionals to program their own learning materials and exercises for their students. Casner, who spent two years working to emulate the program found in an aircraft's flight management computer, was assisted by Smiths Industries, Boeing, the FAA, and several U.S. airlines.

The software program features five windows—a CDU, mode control panel (MCP), two maps showing the aircraft's lateral and vertical tracks, and a flight mode enunciator showing which flight systems are currently controlling the aircraft. There also is a section to the right of the CDU for entering text and carrying on a dialogue with the pilots. The software program is designed to use video, audio, and film clips to illustrate flight sequences.

Casner said that "our goal is to investigate new technologies and techniques" to help meet training challenges as airplane systems become more complex. He said he has demonstrated the program to several major U.S. airlines and has received inquiries from foreign carriers, and he plans to test the program in a major airline's pilot training program. Casner also plans to create a training program for university students who want to become pilots. He hopes to produce a textbook and CD-ROM introducing students to modern flight management systems.[5]

Wish list. If you are considering CBT system or personal flight planning system development as a possible outlet for your software code-writing skills, keep in mind that the best of these programs have one or more of the following characteristics:

- They are produced in CD-ROM format or use compression techniques to minimize installation effort and resources.

- They don't alter the user's computer memory or CONFIG.SYS files.

- They provide expanded briefings and system tutorials that furnish overviews of system design and operation.

- They are available in demo versions of sufficient detail to provide the potential user with a clear sense of what the program offers.

- They are reasonably priced and offer some recompense if the user is unsatisfied with the product.

- They provide free online support for a reasonable period of time (90 days, in most cases), and charge reasonable fees (per call) for support after the grace period.

- They provide a user group message board on an Internet home page or online service area forum (such as CompuServe).

- They provide regular, affordable software updates as necessary and encourage the use of shared-solution upgrades (e.g., a fix requested by one or more users should be available for all users in the next version of the software).

Langley: Sim world

No tour of the simulation industry would be complete without mentioning the NASA Langley Research Center's Simulation Systems Branch Facility in Hampton, Virginia. The facility operates eight flight simulators that are used for a variety of research programs. Among these programs are ACTS, DMS, MOTAS, and TSRV, which are described in the following sections.

ACTS. The Advanced Civil Transport Simulator (ACTS) is a futuristic aircraft cockpit simulator designed for researching issues that will affect future transport aircraft flight crews. The ACTS objective is to heighten the pilot's situational awareness through improved information availability and ease of interpretation to reduce the possibility of missed signals and misinterpreted data. The simulator's five 13-inch CRTs are designed to display flight information in a logical, easy-to-see format.

Two-color, flat-panel CDUs with touch-sensitive screens provide monitoring and modification of aircraft parameters, flight plans, flight computers, and aircraft position. Three visual display units have been installed to provide out-the-window scenes via the simulator's computer-generated image (CGI) system.

The major research objectives of ACTS are to examine needs for transfer of information to and from the flight crew; study the use of advanced controls and displays for all-weather flying; explore ideas for using computers to help the crew in decision-making; and study visual scanning and reach behavior under different conditions with various levels of automation and flight deck-arrangements. ACTS also provides a test-bed to study and exploit advanced automation concepts.

The major ACTS research application is the graphical control display unit (GCDU). The CDU on a modern transport aircraft is used by the pilot as interface to the flight management system (FMS). The graphical CDU experiment is an evaluation of an FMS interface oriented toward reducing pilot input error while increasing pilot understanding of system and flight-path situation. This new interface was developed using modern graphical user-interface techniques, and it is hoped that this design will better support the pilot in interacting with the FMS, as well as improve pilot efficiency while reducing airline training requirements.

DMS. The Differential Maneuvering Simulator (DMS) provides a means of simulating two piloted aircraft operating in a differential mode (the pilots maneuver relative to each other) with a realistic cockpit environment and a wide-angle external visual scene for each of the two pilots. The military is the primary user of this simulator because of its applications to enhance air-to-air and air-to-ground combat techniques.

The DMS consists of two 40-foot-diameter projection spheres. Each sphere contains an identical fixed-based cockpit and projection system. Each projection system consists of two terrain projectors to provide a realistic terrain scene, a target image generator and projector, a laser target projector, and an area-of-interest projector, which project onto the sphere. The terrain scene, driven by a CGI system, provides visual reference in all six degrees of motion, allowing unrestricted aircraft maneuvering.

Each cockpit provides color graphics electronic displays; standard pilot controls such as stick, rudder pedals, and throttles; simulated engine sounds and wind noise; and cockpit vibration, which add realism. In addition, G-suit and G-seat pressurization systems are used to simulate forces of gravity that a pilot experiences while performing maneuvers during flight. This dual simulator can be tied to a third dome (the general-purpose simulator, also located at Langley) and thus provides three aircraft interaction when required.

The DMS allows for research of highly agile aircraft that can operate effectively over a greatly expanded maneuvering envelope; provides a realistic, wide (field-of-view) maneuvering environment and allows for flexibility and repeatability of maneuvering conditions for high AOA (angle-of-attack) research; and provides an efficient simulation cockpit environment in terms of pilot visibility, the display of flight instruments, and the use of a realistic force-feel system for the pilot stick and rudder pedals. DMS research applications include studies of advanced flight control laws, helmet-mounted display concepts, and performance evaluation for new aircraft design concepts for development programs.

A major goal of developing flight dynamics technology for high angles of attack is to provide enhanced agility and handling qualities that will enable future aircraft to perform unconventional maneuvers that can be very advantageous in air combat. These capabilities enhance safety by preventing loss of control in virtually any flight conditions.

One of the major research applications is the High Angle-of-Attack Research Vehicle (HARV). The HARV simulation represents a highly modified F-18 fighter aircraft that is used as a flying test-bed for advance controls research. Studies are focused on flight at high AOA to

- Assess the impact of advanced control concepts on stability and maneuverability and development of design criteria for advanced control systems
- Develop flight control laws to effectively utilize high AOA maneuvering capability
- Improve control system design methods for flight at high angles of attack
- Assess the effect of airframe and engine modifications
- Investigate impact of advanced controls on tactical utility
- Enhance pilot awareness and management of energy through advanced cockpit and helmet-mounted displays
- Develop and evaluate flight test maneuvers useful at high angles of attack

MOTAS. The Mission Oriented Terminal Area Simulator (MOTAS) facility is a simulation of an airport terminal area environment using several aircraft simulators and an air/ground communications network. The airport terminal area represents Denver Stapleton International Airport and surrounding area with either an advanced automated ATC system or a present-day vectoring ATC system using air traffic controllers.

MOTAS combines the use of several aircraft simulators and "pseudopilot" stations to "fly" aircraft in the airport terminal area. The facility is equipped with various simulators that allow flight crews to fly realistic missions in the airport terminal area. The remaining aircraft flying in the airport terminal area are flown through by the pseudopilot stations. The operator of these stations can control five to eight aircraft at a time by entering commands to change airspeed, altitude, and direction. The other major components of MOTAS are the air traffic controller stations, which are configured to display and control the two arrival sectors, the final approach sector, and the tower or departure sectors.

MOTAS provides a flexible and comprehensive simulation for the airborne, ground-based, and communications aspects of the airport terminal area environment. It also allows research of ATC concepts that would not be possible in the real world due to safety, economic, or repeatability considerations. Recent areas of research at MOTAS include controller considerations in aircraft/ATC data link message transfer.

TSRV. Here's a riddle: When is a simulator a real airplane and when is a real airplane a simulator? Answer: When it is the Transport Systems Research Vehicle (TSRV).

The TSRV and TSRV simulator are primary research tools used by the Terminal Area Productivity (TAP) program. The goal of the TAP program is to increase the operational capability of modern aircraft and foster their integration into the evolving national airspace system (NAS).

The TSRV is a highly modified Boeing 737. The airplane is equipped with two flight decks: A conventional Boeing 737 flight deck up front provides operational support and safety backup, and the fully operational research flight deck, positioned in the aircraft cabin, provides the capability to explore innovations in display formats, contents, and in-aircraft operations. The TSRV simulator provides the means for ground-based simulation in support of the TAP research program.

Promising simulation research results become the subjects of actual flight test research. The simulator is fully integrated with a realistic air traffic control facility to provide an environment for systems level studies. The TSRV simulator allows proposed concepts in such areas as guidance and control algorithms, new display techniques, operational procedures, and human/machine interfaces to be thoroughly evaluated.

The TSRV simulator also provides the crew with realistic real-world scenes. Its wide field-of-view display systems are capable of daytime, nighttime, and all ranges of weather effects. It is also equipped with a full complement of electronic flight displays and two side-stick flight controls representative of the technology available in commercial transports of the 1990s.

One of the major TSRV simulator research applications is the Cockpit Weather Information Needs (CWIN) system. CWIN is a graphical weather system that provides national weather radar images, ground lightning strikes, and surface observations on a center display of a cockpit using either live or recorded data. A green-yellow-red color-coding system shows the intensity of the radar echoes, and lightning reports appear as yellow dots. Pilots interact with CWIN via a touch-sensitive screen. Its ultimate purpose is to burn less fuel and fly shorter distances to avoid bad weather. CWIN's major goals are to

- Enhance the dialogue capabilities between pilot crew and dispatcher for reroute decisions

- Provide historical weather information during flight for references

- Help develop meteorological tools for the pilot crew and dispatcher to help cut time and workload required in obtaining real-time weather situational awareness.

Hanging out at the virtual airport

In coming years, computer-based training aids such as flight simulators, FTDs, PC-FTDs, and other variations will be playing an even greater role in aviation and, indeed, might be the springboard for many of the concepts we'll see in future aircraft cockpits.

Many folks in aviation, like myself, got started in this business after spending a lot of time hanging around airports and bumming plane rides. Unfortunately, because of airport security concerns, this is becoming harder to do in many places. That, plus the lure of MTV, Game Boys, and electronic arcade games at the local mall might be diverting a lot of kids from even thinking about taking up flying—either as a career or just recreationally.

But maybe—just maybe—the PC can stem the loss. I don't think I am alone in believing that someday, some kids will hang out at virtual airports on the Internet, and maybe a few of them will move up to a "real" simulator—and one day walk out onto an airport ramp and climb into a real cockpit.

References

1. FAA Draft Advisory Circular 120-45A. 1990. Airplane Flight Training Device Qualification. September 1.
2. Ibid.
3. Gormley, Mal. 1993. "Training Alternatives: Simcom." *Business & Commercial Aviation*. August.
4. Gormley, Mal. 1994. "The Digital Classroom." *Business & Commercial Aviation*. December.
5. *Aviation Daily*. 1996. March 18.

Air Carrier Computing

Air carriers continue to post record bookings on their flights, and they are ordering new aircraft as fast as the manufacturers can produce them. Healthy load factors across the industry and strong yields continue to help buoy operating revenues. In 1995, U.S.-based air carriers earned an estimated $2-billion profit, thanks in no small part to their use of computerized systems for passenger reservations, aircraft and crew scheduling, maintenance management, statistical analysis, and a host of other applications. Computers have enabled airlines to automate many tasks, and new applications are being developed all the time. However, one gets a mixed impression when it comes to the business of airline computing. Some carriers, such as American, United, Delta, British Airways, Singapore, and a few others appear to be on the cutting edge of computer technology and are willing to invest in advanced systems in search of additional profit or market share. Despite this, American, for example, might spin off its SABRE Group, Inc., as the company's parent AMR Corp. looks to focus on its core business.

Many other carriers recognize the value of using computer-based information systems to streamline their operations but turn their computing tasks over to outside computing and network services either because they lack the expertise in-house or they have chosen to concentrate on their primary stock-in-trade: moving people and cargo by air. And some carriers won't discuss their computing capabilities at all, apparently in fear of divulging trade secrets to competitors.

The major airlines have been computerized for quite some time, but in the past three or four years, say industry analysts, a number of carriers have begun to upgrade their computer systems. The primary impetus for this effort has been the carriers' desire to take advantage of the latest generation of faster, more compact, more manageable computers and more compliant software that is available. A second motivation is simply knowing there must be, in the words of an American Airlines software specialist, "a better mousetrap—the drive to improve the bottom line always drives the search for improved methods of doing business."

Major airlines are those earning revenues of $1 billion or more annually in scheduled service. These used to be called trunk carriers, and they generally provide nationwide and, in some cases, worldwide service. In terms of RPMs (revenue passenger miles), the top 10 major airlines of the world in 1994 (the latest year for which data are available) were United, American, Delta, Northwest, British, Japan, USAir, Continental, Lufthansa, and Air France. There were nine major U.S. passenger airlines in 1994: American, America West, Continental, Delta, Northwest, Southwest, TWA, United, and USAir. In addition, two all-cargo airlines were classified as majors: Federal Express and United Parcel Service.

The airlines' primary interface with the world are the computerized reservations systems they depend on. And, it would seem, even this tool appears to be about to undergo a major transformation.

Reservations Systems

Passenger reservations systems are the life-blood of the airlines. Travel agencies play an important role in airline ticket sales. More than 85 percent of the industry's tickets are sold by agents, most of whom use airline-owned computer reservation systems to keep track of schedules and fares, to book reservations, and to print tickets for customers. Airlines pay travel agents a commission for each ticket sold. There are more than 30,000 travel agents in the United States, providing a vast network of retail outlets for air transportation that would be enormously expensive for the airlines to duplicate on their own. Similarly, freight forwarders book the majority of air cargo space. Like travel agents, freight forwarders are an independent sales force for airline services, in their case working for shippers.

The leading airline reservations systems include SABRE, Worldspan, System One, Apollo/Galileo, Galileo/Sierra, and PARS. When I asked airline experts which were the top three most advanced and capable airline reservations systems, the answer I got was invariably, "1. SABRE; 2. SABRE; and 3 . . . SABRE."

The SABRE story

When airline service began, passengers simply showed up at the airport—if the airplane worked, if the pilot showed up, and if the weather was good, there was a fair chance of getting where they wanted to go. Eventually, passengers could call the airport and reserve seats in

Figure 5.1 Airlines are quickly finding that many passengers are comfortable making reservations online. Some reservations systems are planning on shifting their focus away from the traditional travel-agency market, with its high overhead, but don't expect the travel agency industry to just fold its tent and go home. (AMR Corp.)

advance, but the increasing demand for seats and other factors quickly led to a search for a more predictable—and profitable—way of selling seats on airplanes. The search was on for what would someday be called an automated reservations system.

American Airlines' first pioneering effort with reservations was the "request and reply" system used in the 1930s. A reservations agent would telephone the central control point where inventory was maintained to inquire about space available on a flight, and a response would be returned via teletype.

Through the mid-1940s, reservations were recorded manually with a pencil on colored index cards, nicknamed "Tiffany" cards after the famous lamps with the colored glass shades. These cards were arranged in a lazy Susan, and flights were controlled by half a dozen employees sitting around a table spinning the lazy Susan for index cards that would correspond to particular flights. By counting the pencil marks on each card, a clerk at the reservations center could give a "yes" or "no" to a request for a seat.

In some reservations offices, a wall-sized status board was installed to display seat space available on each flight. The board summarized much of the information on the index cards in the lazy Susan. As new reservations came in, workers at the table passed the information to board workers, who removed seats from a particular flight until no seats remained.

Using the Tiffany system to complete a booking for a round-trip reservation from New York City to Buffalo required 12 different people performing more than a dozen separate steps during a three-hour period—longer than the flight itself!

American is credited with developing the industry's first electromechanical device for controlling seat inventory in 1946. It was called the Availability Reservisor, and it applied basic computer-file technology to the task of tracking American's seats and flights. Even though it couldn't sell the seat or cancel a reservation, the system represented a milestone in adapting electronics to airline reservations. By 1952, the airline had added basic computer file technology—a random access memory drum and arithmetic capabilities to the Reservisor. With the Magnetronic Reservisor, a reservations agent could check seat availability and automatically sell or cancel seats on the electronic drum.

As advanced as the Magnetronic Reservisor was for its time, the airline reservations process was still intensely manual. In 1953, a chance meeting of two Mr. Smiths on an American Airlines Los Angeles-to-New York flight resulted in the development of a data-processing system that would create a complete reservation and make all the data available to any location throughout American's system. The outcome of the conversation between C.R. Smith, American Airlines' president, and R. Blair Smith, a senior sales representative for IBM, was the 1959 announcement of a Semi-Automated Business Research Environment—better known today as SABRE.

American's initial research, development, and installation investment in this system was almost $40 million—the price of four Boeing 707s at the time! By 1964, the telecommunications network of the SABRE system extended from coast to coast and from Canada to Mexico. It was the largest real-time data-processing system—second only to the U.S. government's system known as SAGE.

In May 1976, American installed its first SABRE unit in a travel agency. By the end of the decade, SABRE had more than 1000 travel-agency customers. Today, more than 300,000 devices in 74 countries on six continents are connected to SABRE.

What began as a system for American Airlines to keep track of seats sold on its flights has evolved into an electronic travel supermarket used by travel professionals, corporations, and consumers worldwide to book airline, car, and hotel reservations as well as to order theater tickets, bon voyage gifts, flowers, and other travel-related goods and services. The introduction of easySABRE in 1985 allowed personal computer users to tap into SABRE to access air, hotel, and car reservations.

SABRE is used to book an estimated $45 billion in travel products each year. SABRE has evolved into the world's largest privately owned real-time computer network and has processed a record high 4176 messages per second.

The SABRE system is considered by industry observers to be one of the most, if not the most, advanced computing systems in the world. The system is reported to be capable of tracking some 600 aircraft around the world on a minute-by-minute basis worldwide, as well as all the passengers who will fly aboard these aircraft within the next year. At the same time, the system can accept reservations requests from thousands of travel agents at terminals around the world, as well as through its easySABRE network.

How good a system is it? In the words of one SABRE employee, "We have all of the latest stuff from all of the top companies—IBM, Sun, SGI, HP, Auspex, etc. Not only workstations, but huge multiprocessor monster servers as well. We push these systems to the limit with the software our developers create to manage not only AA's operations, but other airlines worldwide.

"In the hallway down from my office are a couple of huge crates from Cray. Care to guess what's in them? We also have a couple of SP2 parallel-processing systems from IBM in addition to their ever-present mainframes. How advanced are we? I think most commercial firms would love to have half of what we work with everyday."

The SABRE reservation system runs on IBM mainframes running the TPF (Transaction Processing Facility) operating system. UNIX servers are reportedly used for add-on services such as fare calculation, decision support, slot allotment, flight operations, cargo handling, and baggage tracking. The database running on the reservation system is based on a package called TPFDB, which is a development of Swissair Information Systems and is now available as a IBM software, TPFDF.

Since SABRE and its parent, AMR Corp., were among the first to develop computing systems for reservations, flight operations, and maintenance tracking, they have managed to command the lion's share of the market.

Web rez

What does the future hold for reservations systems? It's good bet that the Internet will figure prominently in airline travel planning. Any business based on providing the public with information from private online systems, such as stockbrokers or travel agents, will be challenged by direct public access to the same information available on the Internet. One can now plan an entire vacation—including airline, car rental, and hotel reservations, as well as dining, sight-seeing tours, and entertainment planning, by using the Internet. This development is certain to change how airlines book flights, which will in turn will cause changes in all reservation systems. Many airlines are already encouraging passengers to use the airlines' Web sites to plan their itineraries, select seats, peruse menus, and use other services the carriers offer.

Due to emerging distribution technologies (the Internet, mostly), some carriers, such as Delta, have already asked their computer reservations system (CRS) vendors to reduce their costs or face being eliminated altogether. American is seriously toying with shedding its SABRE Group holding for a number of reasons, but competition from new technologies is among them. And USAir and British Airways have developed a system that will allow their customers to book flights using a personal computer. The system, called Priority TravelWorks by USAir and Executive TravelWorks by British Airways, is designed to simplify airline, hotel, and rental car reservations.

Another approach might do away with tickets altogether. TWA has begun limited testing of electronic ticketing through Worldspan and planned to introduce ticketless travel to its customers in 1996. "In cooperation with our airport support services, staff, computer systems experts, and Worldspan, we are starting limited testing in the St. Louis, Kansas City, Columbus, and Indianapolis markets," said Jan Wood, vice president of domestic sales. Thirty TWA employees who travel on business made trial flight reservations through the system before it was opened to the public in mid-April 1996. TWA expected the full domestic system to be available for electronic ticketing by the summer of 1996.[1]

ATPCO

Pricing and scheduling can make or break an airline, and both have become more complicated since deregulation. Airline prices change daily in response to supply and demand and to changes in the prices of competitors. Computers are used to track fares closely, and tens of thousands of air travelers worldwide benefit from the specialized data services of the Airline Tariff Publishing Company (ATPCO) every day of the year.

ATPCO maintains and distributes domestic and international passenger fares and associated rules, as well as cargo rates. The information is distributed to all computer reservations systems used by travel agencies and airlines and is filed with participating government agencies. More than 2.5 million transactions are processed and distributed by ATPCO on an average day.

For this task, ATPCO relies on an IBM ES/9021 mainframe server. It allows ATPCO's staff to create, adjust, and distribute thousands of fares daily. More than 270 airlines worldwide transmit data to ATPCO via data links, tapes, diskettes, or in hard-copy format. Based on this information, ATPCO establishes new fares or adjusts existing ones through the use of complex algorithms. The new and revised fares, as well as automated rules, are distributed worldwide to computer reservation systems one or more times daily. The fares are displayed on the reservation systems and used to automatically price itineraries and issue tickets.

Many countries require that airline fares and rules covering transportation to and from that country be filed with and approved by an agency of their government. Filing was done on paper by the airlines serving those countries until ATPCO developed the industry's first online system for handling the requirement. Electronic filings are now accepted by a growing number of countries, including the United States. Some 25,000 separate filings, involving more than 16 million pieces of information, are stored annually through ATPCO's electronic government filing system. Speed and accuracy are essential in this business. Slow computer response time can be disastrous to client airlines if it prevents them from implementing fare changes on short notice.

ATPCO is owned by 29 North American, European, and Asian airlines and has been serving the airlines' tariff requirements for more than 50 years. It also provides a similar type of support for car rental rates. The company's main office is located near Washington, D.C., and its European office is located near London.

Schedules don't change quite as frequently as fares, but they do change far more often than they used to when the government regulated the industry. The carriers use their computer reservation systems to advertise their own fares and schedules to travel agents and to keep track of the fares and schedules of competitors. Travel agents, who sell the bulk of all airline tickets, use the same systems to book reservations and print tickets for travelers. The carriers also are constantly analyzing their fare structures versus their competitors. Since deregulation, this has been a subject of constant study by insiders and industry observers.

By the way, one of the better titles on deregulation and the future and status of U.S. and international airlines is *Airline Odyssey: The Airline Industry's Turbulent Flight Into the Future* by James Ott and Raymond E. Neidl (McGraw-Hill, 1995).

Airline Data Services

Airlines are among the most data-intensive industries of any business. Although some of the larger carriers perform their own analytical functions, many carriers turn to outside analysts to obtain critical information about industry finances, traffic, fleets, and schedules.

For example, when carriers are considering expanding service to particular markets, they need information about current international and domestic traffic and capacities, fares, quality of service indices, city pair market and yields, and traffic forecasts. One of the larger providers of this kind of information is BACK Information Services. The company provides a wide range of analytical and support services and analytical software tools, including airline performance and outlooks; market economic analyses; route and schedule development; airline feasibility studies; business operating plan development, and other services. The company also provides access to Form 41 data, which consists of financial information and statistics filed with the U.S. Department of Transportation (DOT) by U.S. certificated airlines. This information provides carriers with a look at their competitors' balance sheets, profit and loss statements, traffic schedules, and other data. The company also produces route mapping software so carriers can get clear, visual representations of competitors' markets and equipment activity, making it easier to decide whether to enter a particular market.

Aircraft and Crew-Scheduling Systems

In the early days of airline operations, the process of assigning flight crews to aircraft was pretty straightforward: A carrier's senior pilots flew the company's routes Monday through Friday, while the newer pilots flew on the weekend—period. If you were the dispatcher, and one of the pilots called in sick or quit, you just hired someone else. If you needed more airplanes, you either built them yourself or bought them.

Over the years, however, as the airlines expanded and organized labor began to be a factor and later, as FAA duty time limits, training, and medical certification rules became part of the equation, scheduling aircraft and crews became a lot more complex. Although some major carriers were still scheduling crews and airplanes manually as recently as 1991, most large carriers had transferred most of their basic scheduling processes to computers by 1980. Indeed, a lot of the computer industry's success can be attributed to the airline's use of mainframe computers and computerized networks.

Denver-based United Airlines is typical of the large carriers of the mid-1990s. United operates two large, separate mainframe computers. Its IBM mainframe server is used for its Apollo reservations system, and it relies on a Unisys 2200-900 series mainframe for its flight operations systems. United has even developed enough computer processing and storage redundancy to market the extra capacity to other carriers in need of redundant systems of their own. Response time on the Unisys is under two seconds for most applications, although to compute a complete San Francisco–Hong Kong flight can take as long as 20 seconds.

The United system, like many used by air carriers, is used for planning flight routes, altitudes, navigational fixes, and communications frequencies; creating and forwarding FAA flight plans; calculating aircraft weight and performance; creating aircraft manifests; connecting with various private communications networks; compiling and maintaining flight planning information (weather, NOTAMs, etc.); creating monthly flight crew bids; pairing and scheduling pilots and flight attendants; tracking aircraft maintenance; and many other functions associated with corporate operations.

While there might be some variation from one carrier to another, airline crew management systems generally perform very similar functions because most crew management tasks are common to all carriers. Programs that are used for this function keep department managers and crew schedulers informed of the location and "legal" status of each of the airline's flight crewmembers. Legal in this sense has to do with remaining inside a very rigid set of rules that define a crewmember's qualifications to fly on a given day. Some crewmembers, for example, will be on duty, some will be on reserve status, and others will be off duty. Some crewmembers will have just come off duty and therefore cannot be called back to cover another flight for a specific period of time (called the required rest period) and must not be engaged in any non-flying company duties such as training or paperwork. The Federal Air Regulations (FARs) and company policy manuals also specify:

- The maximum number of hours a crewmember may fly in a given day, week, and month

- How often pilots and flight attendants must receive recurrent training

- How many instrument approaches, landings (daytime and night-time), and hours of instrument flight are needed to be considered legal for recency-of-experience considerations

- When crewmembers must undergo a company physical exam

While all this information can be maintained on a manual basis, rules-based computer tracking systems are often used to automate the process and provide additional information that eases management decision-making.

For all but the smallest carriers, crew scheduling is a very complex, dynamic process, even under normal circumstances. When external factors arise, the schedulers, flight planners, and dispatchers—and their computer programs—are put to the test. For example, when snow-storms repeatedly blanket the country and cause widespread airport closures and delays, as they did in the winter of 1995–1996, carefully prepared crew scheduling plans can fall apart in a hurry as crews divert their aircraft to alternate airports or are otherwise delayed. Reserve crewmembers (who already have been assigned schedules to fly) must be called in to cover the gaps in the schedules. But then the reserve crewmembers' schedules must be adjusted so that all the carriers' crews don't run afoul of FAA, company, and union rules before the month ends. Throw in normal employee absenteeism for illnesses or other personal reasons, and you can imagine the effect of a labor dispute, an FAA-ordered fleet grounding, or a merger with another carrier.

Add to this is the fact that many, if not most, crewmembers move around a lot as their schedules take them from one part of the system to another. Scheduling systems have to able to account for dead-heading time and time away from the home base, as it all counts toward the duty time limits. Furthermore, additional airline personnel must be positioned for ramp, gate, security, and baggage-handling tasks.

Other factors, such as the experience level of each pilot, must be taken into consideration, so that two relatively inexperienced pilots aren't assigned to a flight. Some scheduling systems are now capable of producing an "audit trail" so that if one or more crewmembers' schedules develop unexpected legality problems, managers can determine when and how the problem developed.

Most airline crewmembers are asked to "bid" or request flight assignments according to personal preferences. In a perfect world, the ideal scheduling system would always match everyone's desire to fly certain routes with particular crewmembers on certain days of the week, excluding birthdays, holidays, family events, signs of the Zodiac, and other criteria, but it isn't a perfect world, so schedulers have to make lots of compromises. A capable crew scheduling system can help them immensely.

As this book was being written, new rules to harmonize the flight and duty time requirements of FAR Part 135 and Part 121 carriers were being introduced. Several acquaintances in the business of writing scheduling software were trying to figure out how to make the changeover as smooth as possible for their clients. One quipped, "I just hope [the FAA] has a person on the [Rulemaking] committee that has some input on how the new regs can be computerized. The old regs were written in the '60s, and they're not easy to comply with on a computer."

Another friend related that at one carrier he worked for in the early 1980s, the crew scheduling system was computerized, but other tasks were still manual. Thus, whenever a small change was made to the flight operations routine, crew schedulers would find themselves dealing with new headaches. The carrier eventually computerized its entire operation.

Scheduling a large mixed fleet of aircraft can be equally challenging and requires equally capable software and seasoned schedulers. The FAA's rules for regular, routine aircraft maintenance would be demanding enough, but the real world complicates it more. Parts wear out before they're expected to, the wear-and-tear in some kinds of operations are tougher on airframes than others, and despite efforts to standardize operations, pilots don't always operate aircraft exactly the same.

At the same time, air carrier managements are driven to maximize the time the aircraft are available for flights, while holding down maintenance costs as much as feasible. Additionally, the vagaries of aircraft schedules don't always leave airplanes parked next to the carriers' maintenance hangars at the end of the day, so someone must either taxi or fly the airplane to where it can be serviced.

Computers are also essential in maintaining the carriers' stated personalized service goals. This means, for example, that the galley services they provide must be coordinated with catering services to ensure that the passenger in seat 37F on Flight 1621 gets the dietetic, salt-free, kosher meal that was requested by a travel agent eight months ago.

Flight Planning, Dispatch, and Flight-Following Systems

The capabilities of air carrier flight-planning systems can range from handwritten flight plans to high-speed mainframe computer systems that can calculate fuel- and wind-optimized flight plans to the nth degree. Most computerized systems are operated on mainframe or minicomputers, although some smaller airlines are starting to use systems designed to run on networked PC workstations. Fortunately, the level of computing has advanced somewhat beyond the early 1980s when many single-task systems of the period, such as flight planning and crew scheduling systems, couldn't be easily integrated.

The task of a flight-planning system is to determine the most efficient routes and altitudes to fly under each day's set of weather and operational considerations. Once the calculation is complete, the system provides the flight crew with a flight plan containing all the information they need to fly to their destination (e.g., departure, en route, and arrival fixes and routes; airport information; aircraft weights; and weather reports and then continue on to an alternate airport in the event the airplane can't land at the primary destination. At the same time, airline dispatchers see that the operational plan is sent to the FAA or CAA for processing.

Dispatchers also have to be able to stay in contact with the carrier's flights to provide updates of weather and airport conditions or other operational information, so many now use aircraft situation displays (ASDs) to monitor aircraft flights-in-progress. Communications to and from the aircraft are accomplished via the ARINC/SITA networks. Dispatchers also provide flight releases once the doors on the aircraft are closed and the aircraft's weight and balance and other performance figures have been calculated. Dispatchers also must coordinate with crew schedulers to ensure that flight crews don't exceed their time limits. Most large air carriers have automated most of these tasks.

The FAA and other regulatory agencies are giving some consideration to certifying airline flight-planning systems, but as of this writing, nothing definitive has been proposed. It is conceivable that certification of weight and balance, flight planning, and weather systems used by air carriers might be included in any certification effort. Software developers are also watching to see what effect international airspace modernization efforts, such as FANS and Free Flight (described in detail in Chapter 8), will have on flight planning systems.

Free Flight and flight planning

Two developments are likely to have an impact on airline flight planning: the Boeing 777 and Free Flight. Prior to the appearance of the B-777, flight-planning system programmers could safely create an ellipse on the surface of the earth and then limit the most-appropriate-route searches within the ellipse. However, the B-777 can fly halfway around the world, so now a trip from Dubai to San Francisco could create a dilemma. On one day, for example, the most fuel- and time-efficient route could require that the flight depart to the east—over Asia and the Pacific; but on the very next day, with a slight shift in the winds aloft, a departure to the west—over Europe, the Atlantic, and North America—might be more economical. "This adds a level of complexity to route-search algorithms that did not exist before," said an acquaintance in the business.

Free Flight and Global Positioning System/RNAV (area navigation systems) add a second problem in that almost all airline's flight planning systems use a route database of predefined connections between known points (navaids and airports). With GPS and Free Flight, the optimal route-selection process will require searching paths off airways. The algorithms to accomplish this are much more complex and CPU-intensive. Currently only a handful of flight planning systems are up to speed for Free Flight/RNAV planning, but others are expected to have this capability soon. Several carriers have folks working to get their systems ready for Free Flight.

OPS. Airport performance data helps air carriers determine how much payload and fuel can be loaded onboard an aircraft under varying conditions. Factors affecting this capability include field elevation, temperature, humidity, runway and runway overrun availability, and other criteria. Traditionally, these complex calculations have been done on paper or relied on simple databases. The designers of traditional airport performance formats have always had to make choices between ease-of-use and accuracy. Very often, the result is a data format that provides simple but overly conservative corrections for things like headwind or tailwind, nonstandard barometric settings, and aircraft minimum equipment list (MEL) items. Even when air carriers use automated systems to transmit performance data to flight crews, they might contain unnecessary payload restrictions

One company, Computing Devices International, has developed a laptop computer-based software package to provide an alternative solution. The Onboard Performance System (OPS) software package enables flight crews to perform weight-and-balance analyses, takeoff and landing efficiency calculations, and optimum cruise performance analyses right in the cockpit. The real-time computations are intended to eliminate the sometimes costly conservatism found in traditional airport performance data formats. OPS can also improve engine reliability and reduce maintenance costs. In addition, OPS will maximize takeoff and landing weights under all combinations of environmental airport and aircraft system conditions, for example, resulting in optimized payload and eliminating unnecessary fuel stops. The OPS can also maximize takeoff thrust reduction by computing the highest possible assumed temperatures consistent with company policy.

Jeppesen OnSight. Jeppesen Sandersen, the heavy-hitting navigation database firm (described in detail in Chapter 3), is one of a number a providers of airline support systems. Jepp's system, called OnSight, is a group of UNIX software applications designed for the needs of airline operations. OnSight, developed in cooperation with Sonalysts, includes modules for computer-optimized flight planning, weather analysis, near real-time aircraft situation display, and electronic documents. Modules are available individually or as a fully integrated package. Future modules include Electronic Flight Following/Flow Board, Runway Analysis, and Crew Scheduling, as well as other dispatch/operations-related applications.

Maintenance Management Systems

To a maintenance technician, a transport aircraft is a collection of tens of thousands of parts on a set of very large wheels. Keeping track of, repairing, or replacing all those parts requires the use of advanced inventory and data management systems. For the past 10 to 15 years, airline maintenance facilities have increasingly relied on mainframe- and minicomputer-based computer programs and telecom networks to ensure that the rotable and expendable parts used to repair a specific aircraft are available when and where they're needed. The aviation-industry culture has long supported intercompany part exchanges, which are often done electronically. Commercial parts networks also provide a means for carriers to track down infrequently needed parts on a global basis and thus reduce their parts inventory overhead.

The past 20 years have also seen the development and introduction of computer-based aircraft maintenance manuals, illustrated parts catalogs, wiring diagram manuals, specialty manuals, and other related documents. Recent developments have included the introduction of ATA-categorized, hyperlinked documents, aircraft tail-number-specific information, and synchronous and multimedia display capabilities. The trend is sure to continue.

Researching even a small aircraft's maintenance manual and looking up additional documentation contained in manufacturers' service bulletins and FAA Airworthiness Directives could become an overwhelming task for maintenance technicians. So it's not surprising that CD-ROM and other computer-based maintenance documents have quickly become commonplace sights in maintenance facilities around the world. And while I haven't seen it yet, maintenance database searches are reportedly being done via private intranets.

AMOSS: Server in the sky

Avionics maker Honeywell and America West have begun field-test evaluations of Honeywell's Airline Maintenance and Operations Support System (AMOSS). AMOSS provides real-time, inflight maintenance troubleshooting. The testing began in mid-1996 and was expected to last six months. America West agreed to enter into a field trial of the AMOSS system, which was developed by Honeywell and Computing Devices International, to demonstrate the functionality and to validate the performance of a system designed to address the rising cost of airline maintenance and flight operations. Honeywell has received many inquiries about the system from airlines and business aircraft operators since the new system was first announced in 1995. The system might be equally valuable to OEMs who are beginning to offer life-cycle maintenance support for their customers.

AMOSS promises to save airline operators money by reducing turnaround times at the gate and costly troubleshooting on the ground. Two-way, air-to-ground communication (via ACARS or satcom, for

Figure 5.2 Honeywell and America West have begun field-testing Honeywell's Airline Maintenance and Operations Support System (AMOSS). The system is designed to provide a data-link connection between aircraft maintenance facilities and AMOSS terminal-equipped airline aircraft. (Honeywell)

example) allows AMOSS to connect an aircraft's onboard avionics with ground-based support—flight operations/dispatch, maintenance control, maintenance planning, engineering, and line and hangar shops. In this way, data from an aircraft inflight can be downloaded to ground support for analysis and diagnosis through a unique AMOSS diagnostic algorithm based on models.

As this book was being written, Honeywell was developing a fault-model for the Boeing 757, which will be a computer representation of how the aircraft reacts to certain faults. The fault model helps Honeywell system developers to execute the core diagnostic routines.

Computing Devices, Honeywell's partner in the AMOSS development effort, has been focused on implementing these systems on Douglas aircraft (particularly the onboard maintenance terminal), but it is looking to place its systems aboard Boeing aircraft as well. The onboard maintenance terminal is a 486 laptop computer that collects fault information and provides maintenance technicians with information to correct the fault, rather than using a paper document. The system is also capable of relaying fault information while the airplane is in flight to ground-based maintenance facilities to alert them to problems. This way, the necessary materials and personnel can be readied to meet the airplane as soon as it lands and thus reduce down time.

Cargo Management Systems

Air cargo shipping is one of the fastest growing areas in the industry. Both air cargo and regular passenger carriers are fighting for a share of this profitable market. So it isn't surprising that this area is also attracting the attention of developers of computer-based tracking systems. Air carriers are interested in knowing exactly where shipments are at all times so that they can accurately trace their waybills in the increasingly complex world of manifests and airbill tracking. The current leading technologies in cargo tracking are bar code scanners and radio tracking chips.

Tracking the paperwork is the other side of the business. Methods that provide faster paperwork mean faster payments to the carrier. Carriers need a central database to control the tens of thousands of air-

Figure 5.3 The Air Transport Association's Home Page on the World Wide Web caters to its airline members, but it also provides the public with industry-related information. (ATA)

bills they process each month, so they are turning to software companies with an understanding of the intricacies of air cargo. Examples of software vendors in this segment include Cargo Data Management (CDM) and Teledyne Brown Engineering.

Teledyne's TIE-Net system focuses on converting various cargo messages into a common language and can book space, generate airbills, and query for cargo space availability. CDM has three stand-alone products that perform various air cargo waybill tracking and management functions: Money Track II is an air cargo revenue accounting system, Station Track II is air cargo terminal management program, and Easy Track II is an air cargo tracing and tracking system. All of CDM's programs are written in DataFlex relational database. The company counts among its clients Evergreen International, Arrow Air, and American International Freight.

FIDS and MUFIDS

If you want to create instant chaos in an airport terminal, turn off the computers that feed the FIDS. Passengers and airline personnel alike depend on the ubiquitous, colorful flight information display systems, and it is a competitive, innovative field for software developers. Two vendors—Infax and Com-net—dominate the market but share it with a handful of other vendors.

Companies in this market design, manufacture, install, and maintain the systems they provide to airports. In most cases, one vendor is selected by the airport's management to provide the airport's FIDS, although more than one vendor might be present on an airport that has more than one terminal. In this case, the information system is referred to as a MUFIDS—multiple-user FIDS.

Under normal circumstances, FIDS and MUFIDS are used to display flight numbers, departure/arrival times, destination/origin, a countdown timer, and, depending on the location of the display, a last-bag indicator, which informs the ramp crews when the last bag has been loaded. Other information can be displayed on a ad hoc basis, such as gate changes, weather delays, and other information. The systems pay for themselves in savings from misconnected bags, miscatered meals, and misplaced mail. FIDS run on a full range of computer operating systems, from PCs to mainframes.

In addition to the "simple" task of providing passengers with flight information, FIDS are quickly becoming airport intranet providers. Some carriers are turning to FIDS developers to tie together their assorted passenger information systems and operations control software to provide essential information to passengers, agents, flight crews, and others within the airline organization. New types of infor-

Figure 5.4 Federal Express uses Jouve Data Management's maintenance technical documents software. The system provides airframe technical manuals in CD-ROM format. (Jouve Data)

mational services are being added to FIDS capabilities, so it's an open market to imaginative developers.

TOAST

One particularly good example of new FIDS thinking is at the new $50-million Terminal C at the Cincinnati/Northern Kentucky International airport at Hebron, Kentucky. This terminal is operated by Delta Connection carrier Comair. When the terminal's construction began in 1993, Tim Zeis, director of Comair's customer service planning and development, contacted Jim Dublikar, who heads up Comair's MIS team, and Robert Norvell, president of Com-net Software Specialists, Inc., based in Miamisburg, Ohio. The three developed an impressive customized communications integration package for the terminal that could be a model for others.

Comair's challenge was, to quote Zeis, "Challenging. With 225 arrivals and 225 departures a day, and 42 aircraft in our peak hour, keeping track of who and what is on which airplane and when is a significant challenge." The carrier needed a system that tied the company's off-the-shelf systems together in an efficient common storehouse of useful information.

Dublikar and Norvell worked for months and ultimately created the system that exists today: TOAST (Terminal Operations and Schedule Tracking) system. TOAST, running on a UNIX platform, has automated at least five previously manual tasks and integrated Comair's Cincinnati computer systems for under $250,000 and in less than a year from requirements definition.

TOAST ties together five basic electronic information systems: (1) DeltaMatic flight following (time out/off/on/in, estimated time of arrival/departure, and passenger seat assignments); (2) Comair's FIDS, which is provided by Com-net; (3) WX Air (air carrier weather and dispatch data, such as BOW (basic operating weight), alternate airport, max takeoff weight, and minimum fuel onboard); (4) Dispatch schedule control (flight numbers and crew, destination, and gate assignments); and (5) Comair's flight-following system, which is supported by SBS International of New York, Inc.

TOAST also provides access to Comair's walkway monitors showing flight number, destination city, and departure times, as well as the ramp information display system (RIDS). The function of RIDS is to display information, on illuminated 2-by-8-foot electronic displays mounted on the walls above each gate, to cockpit crews and ramp agents without radios. Pilots and ramp agents can see at a glance the flight number, the number of passengers coming to the gate, and a departure countdown clock. If last-minute passengers are on their way out to the gate, the RIDS display begins flashing, indicating that the crew shouldn't push back yet.

"We're trying to give as much information as possible to the crews and agents—and the passengers who ask for it—so they can make some quality decisions," said Zeis. "In the past that was just impossible. For example, there was no way for the individual working in maintenance who was dealing with all these issues, to tell everybody what was going on." Com-net's FIDS could only display so much information. RIDS has been programmed to display a variety of messages. The end result is less "he-said-she-said" workload for the operations personnel in the tower that overlooks Comair's terminal. Instead of having to relay countless routine messages, specialists can now spend more time handling the exceptions.

The Com-net passenger information display system, installed at terminal gates and other locations in operations, provides agents with 64 pages of informational displays, five pages of aircraft mechanical information, five pages of daily station bulletins, a tower coordinator's page, and an employee information page. The system is password-protected.

"One of our biggest challenges was communicating with employees out on the flight line," said Zeis. "For example, before TOAST, dispatch releases for the pilots of our Part 121 airplanes had to have someone [in operations control] print up all the releases. In a 40-aircraft push, there could be as many as 30 of those, each of them 14 to 15 pages long, that were printed in operations control. Then they had to be driven out in a van and delivered to each airplane. And that was assuming there were no changes to the releases. Fuel requests were strictly manual."

To alleviate this situation, Com-net installed 24 walkway workstations in the Comair terminal. The workstations comprise a monochrome computer terminal with a graphical interface, two monitors showing different information, and a dot-matrix printer. Now, flight crews can pick up dispatch materials, order fuel, and even bid on the next month's flights on their way out to the gates.

Another part of the network uses interface devices called "tugmen" located on the outside of the gates. Ramp personnel use these to request or provide information about arriving and departing flights. The information, in turn, is used to update Delta's digital access system for flight information, allowing Delta customers who call the system to receive accurate arrival and departure times. By hitting the IN and OUT buttons, ramp crews also update the flight crewmembers' block and duty time records. "This solves the issue of 'how much you earn depends on whose watch is telling the time,' " said Zeis. "It doesn't matter what time your $700 Breitling [watch] says it is when the tugman clocks you in and out."

Ramp personnel can call up a flight's dispatch release from a tugman to ascertain how much fuel will be required. With fueling stands at every gate (one dedicated to every two airplanes), there is never a need to wait for a fuel truck. The ramp agent only has to check the tugman for the correct flight number, correct aircraft number, and amount of fuel requested. "Now, instead of having 225 fuel slips a day to process, we have only 27, which is the number of fuel carts we have," said Zeis.

Future upgrades call for the development of a paperless manifest, which will, it is hoped, reduce paperwork and operational clutter. More advantages to using the system are being discovered all the time, and the employees are enthusiastic supporters of the system.

Computer learning curve

Despite all the advances in computer technology, it is surprising how recently some of the major carriers still depended on manual grease-board methods. Peter Wanner, who is a Toronto-based independent consultant in the airline software business, recently described the flight-planning center that was used by the "old" PanAm:

"During the period of my involvement with [a flight planning system vendor], they began selling systems to the majors. I was asked to accompany the sales team to PanAm in JFK. The need for the new flight-planning software arose because PanAm's system was so old that the last person that could [maintain] the software had died.

"Being given the tour of the [operations control facility], I was amazed that in 1991, *any* carrier of any size could be so computer illiterate. The [flight planning and dispatch] room, which easily measured 120 by 200 feet, was [filled] with drafting tables. Over the tables were acetate overlays. One wall was devoted to flight planning and the other two walls were used for crew scheduling. The money that was [wasted] away on useless manpower, lost crews, ineffective routings—whew! Who would have believed this!?"

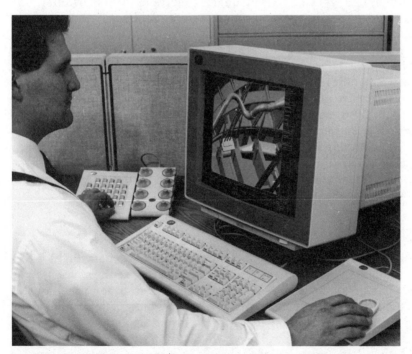

Figure 5.5 Increasingly, airframe manufacturers are asking airline personnel to play a role in the design of the aircraft they specify. (Boeing)

Wanner explained that one carrier he had worked for once completely lost track of one its flight crews during a layover in Athens. Nobody could tell whether the crew had dead-headed home, gone to a hotel, or picked up another flight the next morning. After about a week, one of the pilots was discovered at a hotel used by the carrier. "I've seen that happen three or four times for different airlines," said Wanner. "That's usually when people start thinking 'Maybe we should do something with a computer to avoid this kind of thing.'"

According to industry observers like Wanner and others, one of the big frustrations in the industry is that many systems don't perform all the functions as smoothly as they could, nor will they easily integrate with other systems an operator might need to use. Even the mighty SABRE and other popular programs are not without shortcomings, and some individuals have made a living by pointing out to second- and third-level carriers where their software is working against them and then correcting the problems. Others, like Wanner, work toward integrating systems for their clients' stand-alone systems.

The holy grail of airline information technology managers is a crew-scheduling system with artificial intelligence. AI systems—if they ever become feasible—would have the capability to understand natural language and reason under conditions of uncertainty, and thus could solve the myriad human-factors problems in managing an airline.

Some industry insiders maintain that a surprising number of air carriers' level of computer technology is "way behind" some other industries and even other segments of the aviation industry. Several airline software analysts contend that some major carriers are suffering from "analysis paralysis," due to an overabundance of mid-level MIS specialists and a lack of willingness by senior management to overhaul outdated computer practices. "If you want to make any changes at all, it takes a committee's approval and a team of programmers to do it—and they'll *still* forget to do something to make it work the way it should," said one analyst.

Employment

It used to be that when a carrier wanted to hire new pilots, flight attendants, maintenance personnel, and other employees, the carrier would run a small ad in the local newspaper and wait for the industry grapevine system to bring in thousands of resumes. Then would begin a long process of screening, testing, interviewing, and more screening until the pile of resumes had been whittled down to a class size and the hiring was done—until another class of new-hires was needed. It was a pretty harrowing process for applicants and airlines alike.

Private airline resume services like the Future Aviation Professionals Association (FAPA) improved the situation for a while, but the process was still time-consuming.

Enter the Internet-based employment service. Now pilots and others can, for an annual subscription fee, put their resume on file with a handful of organizations that are providing this service. The applicants can update their resumes whenever it is convenient, and the airlines can screen for applicants by filtering through the service's database. Paperwork is kept to a minimum, and all the applicants need to worry about is the all-important interview and testing.

SABRE school

In another industry-related development, SABRE Travel Information Network has developed a new computer-based, interactive training system. SABRE Personal Trainer, which sells for about $150 and runs on a 486 Windows PC with a sound card and a CD-ROM drive, was developed to make it easier for travel agent subscribers to learn to use the SABRE reservations system. The three-volume CD-ROM set promises to make CRS training easier and more accessible.

SABRE subscribers today travel either to the American Airlines Training Conference Center in Dallas/Fort Worth or to regional centers throughout the United States for a five-day basic, intermediate, or advanced course in SABRE. With the new program, SABRE subscribers have the option of training in or out of their offices, eliminating travel expenses and time away from the office.

The courses include 14 areas of instruction, covering more than 70 different lessons, with everything from ticketing to specialized assistance in marketing rental cars, making hotel reservations, and more. A student can customize the entire courseware by testing out of familiar topics and continuing on to more challenging areas. This feature allows students to design a specific, personal training program. The program incorporates color graphics and audio to make learning more interesting. One of the features of the program is its portability—the user does not need to be connected to SABRE, so lessons can be taken at home on a personal computer; the benefit for users is that they can complete the lessons at their own pace, then practice what they learn on the job. The audio portions of the course are available in English, Spanish, and Portuguese.

Reference

1. *Aviation Daily.* 1996. March 14.

6

Military Aviation Computing: The Challenge of Information Warfare

Computers have played a role in military aviation almost from the outset of the computer age during World War II. Indeed, some of the very first applications to require computers were naval artillery ballistics calculations, which would later have aviation applications as well. As you'll see, the ballistics problems that used to be solved so effectively by the Norden bomb sight and the earliest mainframe computers are now being performed on a desktop computer, and the output is stored on a floppy disk.

The military services have depended on computers ever since WWII, in one form or another, but like the civil sector, the greatest advances didn't take place until the appearance of minicomputers in the mid-1970s. Since then, tasks such as intelligence gathering and archiving, mission planning, reconnaissance evaluation, logistics, battlefield depiction and coordination, damage assessment, and war gaming have been the primary military aviation applications.

Indeed, the military's focus is now on what some observers of the military scene are calling "the fifth dimension" of warfare. They argue, correctly, that until aircraft appeared, in World War I, warfare was fought in two dimensions; by World War II air power's third dimension had changed warfare; with the launch of the first satellites in the late 1950s, warfare entered the fourth dimension—space; and now, in the 1990s, we have entered the fifth dimension—information warfare.

The Fifth Dimension

An acquaintance recently pointed out that the Iraqi air force was, until the first few weeks of the Gulf War, considered to be the fourth-largest

Figures 6.1 and 6.2   Along with greater reliance on small, stealthy unmanned aerial vehicles (UAVs) for reconnaissance, such as Lockheed's DarkStar (top) or Scaled Composites' Raptor (bottom), military commanders are looking to improve or replace their legacy data collection and management systems. (Top: Lockheed Martin; bottom: Scaled Composites)

air power in the world. Was it Saddam's lack of an effective, computer and satellite-based information infrastructure that led to the collapse of his military forces? Many think so.

Yet the average Gulf War air combat unit didn't have access to a significantly greater amount of information than was available in Vietnam or Korea. It still took about two to three days for air tasking orders to be conceived, approved, coordinated, and distributed to the field.

So in the years since the end of hostilities in the Gulf, air force and naval commanders have sought to develop new methods to accelerate the information flow, with an ultimate goal of achieving a near-real-time data flow from the field to command to the field. The task is similar to the private sector's effort to reengineer itself to be more responsive to market changes by leveraging information technology. The effort to institute the fifth dimension is only in its earliest stages, so many opportunities are probably yet to be discovered.

Like many organizations, a cash-strapped Pentagon is looking for mature, robust, off-the-shelf solutions that have been tested in the commercial sector. Along with greater reliance on stealth technology and small, stealthy, unmanned aerial vehicles (UAVs) for reconnaissance, the military is likely to look to improve or replace its legacy data collection and communication resources, as well as study secure, high-bandwidth, private communications networks and satellite links that will be the "weapons" in this new battlefield.

Traditional defense-sector companies such as Loral, Unisys, Logicom, Raytheon, Lockheed-Sanders, TRW, and others are teaming or working alone to develop integrated concepts and projects to meet this new need. In previous chapters I have discussed other aspects of military aviation computing. In this chapter I take a look at some of today's computerized tools of the trade used by the crews of combat aircraft, as well as other topics that provide a sense of the scope of military aviation computing.

Basic Mission Planning and Dispatch

Bob Nagy's 20 years of experience as an aerial tactician, both in the cockpit and in information management, make him well qualified to provide a glimpse at some of the kinds of aviation computing systems used by personnel on the flight lines of U.S. military aviation. Nagy now works as an analyst for a defense contractor in Florida, but in the Air Force Nagy was a major with 1600 hours of flight time, mostly as a navigator/weapon systems officer on F-4E/Gs. He now holds an FAA aircraft dispatcher license and a private pilot license, and when I caught up with him, he was completing his commercial pilot training.

Nagy is also a regular visitor to the Aviation Week Group Forum on CompuServe, which is how I met him. When I contacted him by tele-

Figure 6.3 Scaled Composites Freewing Scorpion UAV, which weighs less than 300 pounds and features a tilting body, was designed with CATIA, CFD, and Ashlar's Vellum 3D software. The military, NASA, and other research agencies are developing UAVs for a variety of reconnaissance missions. (Scaled Composites)

phone, he was in the midst of moving from Texas to his new job in Florida, but he was kind enough to spend a while describing the kinds of computing systems he worked with in his various capacities in the Air Force. Rather than trying to interpret what Nagy told me, I have simply transcribed much of that interview. His laid-back, matter-of-fact perspective cuts through the typical marketing department/public affairs office spin on the military's software capabilities, which, it turns out, are impressive anyway.

When he first became an officer in the late 1970s, Nagy told me the use of computers by line flight crewmembers like himself, "was basically zero. My use of computers came later—in fighter operations, for tactical mission planning, computer simulation gaming, weapons effects analysis, air tasking orders—that sort of thing."

From his descriptions of the operational use of these kinds of systems, one can infer the computing capabilities that made it all possible. (I did, however, caution Nagy not to impart any trade secrets to me.) One of Nagy's areas of expertise in the 1980s and early 1990s was in tactical mission planning. This task involved traditional flight planning combined with threat analysis and ordinance planning.

"Before we'd fly a mission we'd go into the planning room and plot out on the computer what were going to do on each mission," said

Figure 6.4 Dassault's Mirage 2000 jet fighter was developed with CATIA software. (IBM)

Nagy. "I don't recall what brand of computers we used, but basically, we had a room set up with about eight personal computers on a network connected to a database. I forget what the procedure was called, but we'd put these DTMs—digital transfer modules, which were floppy disks (only four times as thick—into the computer drive and download the data onto the disks. We'd download the digital maps of the flight route.

"Based on what was in the computer, you could chart out what you needed to stay away from—the threats [SAM sites and other anti-aircraft weapons]. While you were there you could put in avoidance points so you'd stay away from the threats. The system knew the radius and the range of the threats, so it would chart out a circle you wouldn't want to fly through. We also could find out how low or high we could fly based on the line-of-sight range of an enemy radar. I don't want to get into that too deeply, though."

Into the cockpit

In the cockpit, the avoidance points appeared on a display, and the airplane would automatically turn to remain clear of the hazards. "It's like an autopilot, actually. It would push you away from the danger.

"Then we'd take these disks and upload them into the computer system on our F-4E [Phantoms]. This particular model of Phantom had been upgraded with the ARM-101 DMAS (Digital Modular Avionics System), which was an integrated bombing/navigation system. When it came to bombing, it was accurate as an F-16.

"You would just take this DTM and put it into the airplane's computer and hit the "G" for GO switch, and it would suck everything into the computer—that's all it took. You just had to go through the flight plan and check that your destinations, IP [initial point], target(s), and avoidance areas were correctly loaded. You could confirm that your heading and distance from the IP to the target was just what you had computed in the planning room, for example. That's a warm fuzzy, you know what I mean?

"Back in the planning stage, if you knew your ordnance, you could enter a program that would figure out all the ballistics for you. So when you were on a bombing run, the computer would automatically release the bombs at the exact point for you. You wouldn't have to worry too much about airspeed, dive angle, and all that kind of stuff. It could check out all the ballistics for you. For example, if you wanted to drop six Mark 82s—general-purpose, 500-pound bombs—it would take into account the known winds, density altitude, and other factors; the ballistics computer will release them automatically.

"In the cockpit you'd have a CCIP (constantly computing impact point)—it's just like what you've probably seen on the HUDs [head-up displays] of the F-16. You get a little circle. And when that's over the target, you just hit the BOMB button. When the computer sees the ballistics are right for the ordnance, it'll drop them. It has a certain circular area of probability."

Nagy was quite taken with the capability of the onboard computer on the F-4, and often compared it to the more contemporary F-16 Falcon.

"You could actually be banking and it would get it right. It was really good if you were doing a dive-bombing run, and you were coming down the chute, and somebody was shooting at you, and you were jinking around—with triple-A [anti-aircraft arms] coming up at you. You could still get the bomb off with pretty good accuracy. So if you were dropping six bombs or four CBUs [cluster bomb units], it's good enough."

Once airborne, the aircraft's computer is responsible for providing time-on-target (TOT) information, said Nagy.

"You want to be at your TOT at a certain time. As you move the throttles, you have a digital readout that tells you what time you'll be at the next point on your flight plan."

As Nagy explained it, it is also possible to set the throttles to make sure the airplane would be at the target at a very specific time. By carefully nudging the throttles, the airplane could be kept on schedule. Nagy recounted one mission that required traveling about four hours to arrive at the target, a military bombing range in Puerto Rico.

"We had 12 airplanes—three four-ship [groups]. The TOTs were for 0725, 0730, and 0735, respectively. The first four-ship was 20 seconds early, the second four-ship was 10 seconds late, and the third four-ship was 15 seconds early. So you can see how accurate you can be! And we had the Navy and the Puerto Rican Air National Guard tapping on us as the aggressors, there, too. Boom! We came off the target and jumped on the [refueling] tankers and they took us home."

The program can also be used to set up a air-to-air refueling rendezvous.

War games

Computer simulation gaming is another application of computers that Nagy was involved with in staff positions in Korea. The program was used whenever command post exercises, which were essentially a computer simulation of an air battle, were being conducted.

"We'd input strike packages into a computer by entering the type of airplane, type of ordnance, takeoff times, and targets. Another computer in another location would take in all this information and generate these graphic displays of the missions—as they unfolded—on large map-screens we could watch and simulate that they actually went in and hit these targets. You'd get a dice-roll on battle damage, threats, and loses. Another team would play the Red Team enemy airplanes.

"I've also done this back in the States, at Fort Hood, as an air liaison officer, but it was integrated with the army's ground maneuvers. They were doing the same thing with their tanks. I was working with the Joint Forces Air Component Commander then.

"We'd launch CAS (close air support) missions to help the army with their ground battle. We'd simulate forward air controllers calling in the missions to the Air Support Operations Center. We'd get a TOT back from them and we'd put it into the computer. Then we'd "fly" the mission—you could see this on the air battle screen. It was all coordinated with the army so there wouldn't be any friendly fire.

"The computer would generate a random BDA [bomb damage assessment] report—usually lower than what it would be in real life. I think it was a system run by Logicon, Inc., and I worked a lot with civilian analysts on this project. Sometimes their people wouldn't know what a F-16 was, and I'd give them the information so they could update their computer models of the airplane—for the right performance and ordnance—that sort of thing."

Nagy explained that war-game simulations rely on extensive aircraft and ordnance performance databases, and each branch integrates its data models with other branches.

"So in a battle simulation you'll have two or three models working together. It gives everyone good training in C3I (command, control, communications, and intelligence)." (Recently, observers of the military scene have added a fourth "C" to this term—computers—to make C4I, to evince the increased emphasis on the role of computing in military operations.)

JMEMs. Weapons effects analysis was another function supported by computer applications. The Joint Munitions Effectiveness Manual (JMEM) is a classified document on CD-ROM that comprises the contents of a 40-volume document that describes, in fine detail, the damage capabilities of just about any type of munitions made, from light arms to heavy artillery to airborne tactical weapons, including air-to-air and air-to-ground missiles and "smart" laser-guided bombs. In short, JMEMs is a cookbook for creating destruction and mayhem, which is what the military does best if diplomacy fails.

"Suppose, for example," said Nagy, "that the target is a bridge. I'd use JMEMs to tell me what's the most effective and economical way to take out a bridge?"

The end user can enter certain variables, such as the type of aircraft and ordnance that are available for a combat mission. The program

Figure 6.5   Another view of Lockheed's DarkStar. (IBM)

then produces a probability of success of the mission, based on the number of aircraft and bombs used, for example.

"But if the first answer that comes back isn't acceptable," said Nagy, "such as 16 airplanes using conventional weapons, I can enter a different aircraft, such as an F-15 Strike Eagle using a laser-guided bomb. It might then come back and suggest that only four airplanes are required to take out the bridge. And instead of using 80 bombs, we could probably do it with four precision weapons. That way you can improve your economy of force."

The idea is to match the aircraft to the mission, such as using Maverick missiles to destroy tanks or cluster munitions against an area target.

Cosmic perspective. "I don't think they had this stuff in Vietnam," said Nagy. "I think back then it was manual—you had to crunch the numbers with a pencil. But I think it was used in Grenada, and Panama, and the Gulf War.

"The air crewmember normally doesn't have to do this stuff himself. It is usually done at the staff level, when the air tasking order comes out. I was chief of interdiction plans when I was in Korea, so I was putting together all the strike packages for the air tasking order and all the exercises. I would integrate with all the intelligence guys who had all the munitions-effectiveness software, and I had all the data on what we had available for use—which airplanes, squadrons, how many, the turn times, etc. I'd put all this in the computer and then match it up against the effectiveness to get the most effect out of the available airplanes and ordnance.

"When it comes down to air-to-air, it doesn't matter much—an F-15 is going to carry four AIM-7s or four AMRAAMs and four [heat-seeking] missiles." At that point, it's up to the crews on the AWACs (Airborne Warning and Control) aircraft ("Those are those airplanes with the big Frisbee on the roof," said Nagy) to point out when and where adversaries are taking off and approaching friendly airborne forces.

Nagy holds the Air Force's combat aviation computing systems in high regard. "I never had any disappointments with any of it. If we needed something, we knew it was being worked on, it was in the channels, coming our way."

A follow-on program to DTMs, called Mission Planning System (MPS) was being implemented in the field for F-15 and F-16 squadrons as Nagy was retiring. "I hear that the MPS-1 and MPS-2 are supposed to be real cosmic—for doing almost the same thing as what I had, but they integrate intelligence assets into the mission-planning function."

Other operational systems being used include systems that compute the dive angles, slant range, time-of-fall, release altitudes, speeds, bomb-sight parameters, etc., required to successfully complete a bombing mission and programs to keep track of bomb scores for each airplane, not only for crew evaluation, but for accuracy and maintenance purposes.

Software used for long-range bombers is likely to be similar, Nagy believes, but probably of a higher order, especially for the B-1 and B-2. The crews of these aircraft are likely to be using software that was developed for those specific aircraft. "But the B-52 is a pretty cosmic airplane, too," said Nagy.

I asked Nagy how he'd rate the Air Force's sophisticated weather distribution system, the AFGWC (Air Force Global Weather Center). "I know what it is," he replied. "We had as much weather information as we wanted. We used to have in our squadron a relay from the weather shop for the local weather and so forth, but . . . we'd kinda just look outside—if it looked good, it was good enough for me.

"Of course, if the weather was crappy you'd check to see what the ceiling and visibility was and look at the forecast to see whether you can get back into the field, and you'd pick an alternate. In some squadrons they were getting color weather radar [NEXRAD images] and lightning data. For me, it was usually too much information—more than I need. Maybe the cargo guys need more information, I don't know. . . .

"If there was some kind of super storm, we normally wouldn't fly anyway. But I saw some terrible weather in the Philippines—amazing stuff, really. I was at Johnson Air Force Base on one end of the runway, and at the other end there was a tornado. It was kind of weird. Needless to say, we didn't take off."

Nagy maintains an interest in the subject of mission planning, even as he settles into his new career as a military analyst. He has even been known to play an occasional game of Falcon, which is an F-16 military air-combat simulator system, with acquaintances over a modem line. The program even provides a simulated AWACs ship to point out adversary aircraft and other threats. Nagy describes the game as very realistic. "It's pretty cosmic stuff."

Using laptops for intelligence. Pilots flying missions in Bosnia are taking laptop computers onboard their aircraft so they can get easy access to national intelligence data, according to Marine Corps Captain Brian Schmanske, deputy director of a National Reconnaissance Office division that monitors integration of national intelligence data receivers into weapons platforms. In trying to get around the time and cost problems associated with integrating intelligence data receivers in the cockpit, Schmanske said the pilots are using laptops to get reconnaissance and surveillance updates. While pilots would like integrated cockpit capabilities, Schmanske said progress is slow on most aircraft platforms. "The B-2 [and] AV-8B . . . are in the same boat," he said.[1]

Military air traffic control systems

There is no separate U.S. military air traffic control per se. When military aircraft are operating in the continental United States (CONUS), Hawaii, Alaska, and the possessions and flying in Class A, B, C, D, or E airspace, military aircraft must abide by the FARs as well as each military branch's applicable instructions. Unless they are on an authorized Military Training Route (MTR), within a restricted area, warning area, military operating area (MOA), or proceeding VFR without flight-following, military aircraft are just like any other aircraft on an IFR flight plan.

When training within a MOA, military aircraft separation and coordination is maintained either by an airborne aircraft like an E-2C or an E-3 or a ground-based system such as the U.S. Marine Corp's Air Combat Maneuvering Range off San Diego or at Fallon, Nevada. After exiting a MOA, aircraft normally contact the local civilian approach control if proceeding to or near a civil airport, like NAS Miramar; if proceeding to a military base, they'll contact a RAPCON, such as at Whidbey Island, Washington.

In combat areas, other procedures might be used that call for satellite communications.

It's a small, smart world

Mission-planning tasks haven't changed too much over the years, but the computers that are used to do it are becoming smaller, faster, and even more accurate. The military aviation world of the future is likely to be even smaller and smarter than today's if military strategists are correct in their assessment of information technology's increased importance.

Reference

1. *Aerospace Daily*. 1996. April 9.

Chapter

7

Business and General Aviation Computing

Depending on your aviation background, business aviation and general aviation can be viewed as either one or two distinct entities. The same can be said about the aviation computing products and services used by each of these segments. Business aviation and general aviation computing products span a wide range of functions and capabilities that have been slowly converging as each segment evolves. If you are not part of either segment, it's understandable to lump "bizjets" and general aviation into one amorphous category, distinct from air carriers, military, and utility aircraft (and I do in this book where it seems appropriate). And if you are a participant in either of these segments, you already understand that these areas often have similar and overlapping needs for information and information-processing tools, but some applications are unique to business aviation. So, to make each of these segments' use of aviation computing a little easier to comprehend, we have to stick with some imprecise definitions. As we explore the computing tools used by business aviation and general aviation (separately and collectively), their subtleties will hopefully become more apparent.

Business aircraft can range from light, single-engine airplanes and helicopters to heavy, long-range turbojet transports normally associated with air-carrier operations. Business aircraft are regularly used as a primary means of transportation for company personnel, government VIPs, entertainers, sports figures, and teams. Business aircraft also have been frequently used for carrying passengers for charitable airlift functions (such as participants in the Special Olympics or cancer patients who need treatment at distant medical facilities). Some multifunction business aircraft are also employed to carry freight; transport oil rig crews, technicians, and medical personnel; and conduct research.

Business aircraft can be completely owned, fractionally owned, leased, rented, chartered, or shared, but the primary use is for business-related transportation (although the term "business-related" has been often scrutinized very closely by governmental tax authorities and is an ongoing legal issue). For our purposes, the majority of business aviation operators, however, use turbine-powered, cabin-class airplanes and helicopters and normally have dedicated flight departments made of company personnel or use the services of a third-party flight department management firm. In the United States, most business aircraft are operated under FAR Part 91, but many business aircraft operators are certificated to meet FAA commercial passenger-carrying requirements (spelled out in FAR Parts 121, 125, 127, 129, and 135) and often conduct for-hire operations as an adjunct to their primary missions.

The general aviation category includes any kind of aircraft, from car-top-carried ultralights to gliders, homebuilts, light- and medium-piston and turbine-powered aircraft, antiques, and original or replica warbirds. The category includes airplanes (land, sea, and amphibious), helicopters, gyrocopters, and even hot air balloons and blimps. General aviation aircraft are primarily operated for pleasure, recreation, training, charitable functions (similar to those of business aircraft), or in public airshows. They sometimes carry passengers for business purposes, but unless specifically certificated, they cannot do so for hire.

Another, smaller subset of both segments includes those aircraft used primarily for utility purposes, such as agriculture applications, law enforcement, pipeline patrol, offshore oil platform, research, medevac, and firefighting operations. The operators of these aircraft are as diverse as the aircraft—and the kinds of computing products—they use.

Shared Computing Needs and Capabilities

As noted at the outset of this chapter, both the business aviation and general aviation segments share similar needs for information and information processing. These shared needs include:

- *Flight planning.* The most prevalent use of computing applications common to both segments is flight planning, which encompasses a variety of tasks:

  1. Gathering weather and other operational information for pre-flight planning
  2. Computing an optimal navigational route and altitudes, leg and trip time, speed, distance, and fuel requirements

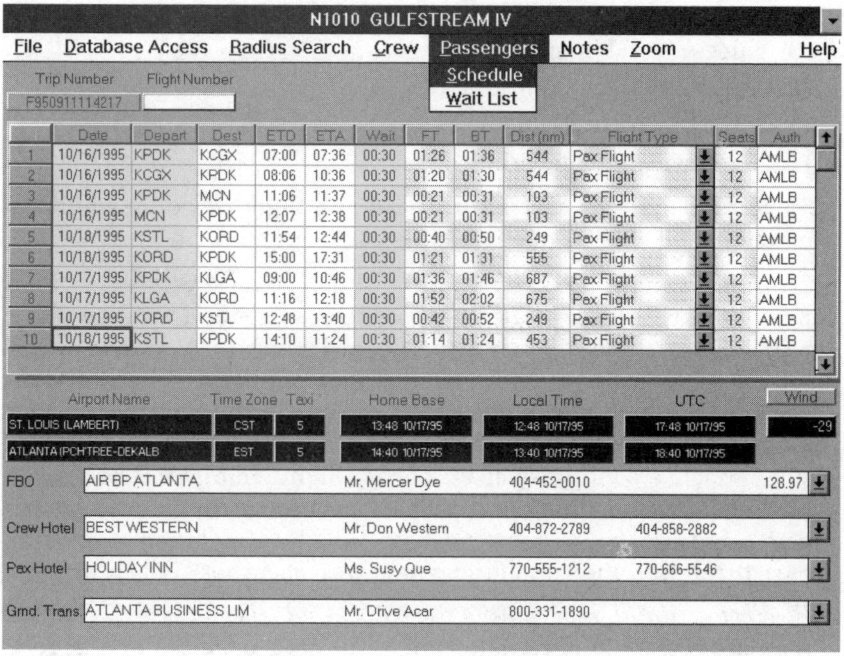

Figure 7.1 Business aviation's need for sophisticated flight department scheduling, budgeting, and maintenance-tracking systems has created a small industry. Currently, about half a dozen vendors cater to the needs of about 4000 operators who require integrated systems like SeaGil Software's BART tracking system. (SeaGil Software)

3. Weight-and-balance calculations

4. Power schedules

5. Oxygen requirements

Many of these programs also keep track of the flight-department pilots' flight times to ensure they meet FAA/CAA recency-of-experience requirements and provide warnings if a compliance date is imminent.

- *FBO management.* Operators of FBOs (fixed-base operators) that cater to business and general aviation clientele typically use computerized systems designed specifically for:

  1. FBO customer billing

  2. Maintenance and repair invoicing

  3. Order entry

  4. Work orders

  5. Inventory management

  6. Part number databases

  7. Fuel tracking

8. Tax tracking

9. Sales-contract management

10. Aircraft tie-down (parking) assignments

11. Personnel record-keeping

12. Accounting and general ledger

Some FBO systems also feature staff and flight scheduling and dispatching and charter quote modules.

- *Training, simulation, and recreation.* Pilots of most cabin-class business aircraft, as well as a growing number of general aviation pilots, regularly attend refresher training programs that include use of highly realistic (computer-guided) flight simulators. Recently, the use of computer-based multimedia systems in the classroom is augmenting, if not replacing, the use of overhead and slide projectors. The immense popularity of PC-based "flight simulators" has generated magazines, online service discussion groups, Internet Web sites, and conventions. We take a closer look at simulation, computer-based training, and simulation gameware elsewhere in this book.

- *Electronic databases, online services, and the Internet.* Another computer-based application common to both business and general aviation is the use of electronic databases, online services, and the Internet. Although we devote an entire chapter to these subjects, they deserve mention in this context. Computer-based information resources are probably the fastest growing element in the aviation industry, and business and general aviation are among the major driving forces behind it, mostly because they are so diverse and fragmented in their informational needs.

  Aviation is one of the more regulation-driven industries, and many activities in aviation require more than passing reference to regulatory, advisory, maintenance, safety-related documents, and other data-intensive materials. A computer equipped with a CD-ROM drive and a modem is quickly becoming one of the more common appliances in the business and general aviation world. Documents and other materials that have been converted from paper documents into shiny plastic discs—and saved a forest in the process—include

1. Electronic FARs

2. Airworthiness Directives

3. Advisory Circulars

4. Aircraft maintenance manuals

5. Aircraft sales materials

6. Listings of aircraft, pilots, maintenance technicians, aviation medical examiners, accidents, and incidents

7. Crew resource management programs

8. Association directories

Appendix A lists sources of aviation-related CD-ROM materials.

For the past 15 to 20 years, many people have found that computer BBSs (bulletin board systems) are a simple, convenient means of obtaining and exchanging information with others. Although BBSs have been overshadowed recently by the Internet, they should be considered a valuable "secondary road" that criss-crosses the information highway. A number of aviation organizations, manufacturers, and government aviation agencies depend on their BBS outlets to communicate with members, customers, and the public about aviation matters. All that's needed to use a BBS is a basic personal computer or dumb computer terminal, a modem, and a simple communication software package.

But if business and general aviation wanted information with just a little more glitz than a BBS, it is finding it in the fast lane of the information highway—the Internet and the even-flashier World Wide Web. Many thousands of pilots, maintenance personnel, engineers, airframe manufacturers, and others in business and general aviation have signed up to online service providers such as America Online or CompuServe or are venturing out directly into cyberspace with Internet browsers. What they're finding is a bewildering assortment of services, associations, newsgroups, archived information, electronic publications (including *AOPA Pilot, Aviation Week & Space Technology, Business & Commercial Aviation,* and *Flying*), government and university resources, weather and flight-planning providers, the National Air & Space Museum, aircraft part exchanges, information about new and used aircraft, business aviation and general aviation-related associations, including the AOPA (Aircraft Owners and Pilots Association), CAA (Corporate Aviation Association), NBAA (National Business Aircraft Association), PAMA (Professional Aviation Maintenance Association), and the home page or e-mail address of just about every aviation-related company you can imagine. An Internet URL (universal resource locator) or e-mail address is included with many of the resources listed in Appendix A.

A Cornerless Market

I cannot overemphasize that the computer programs used by business aviation and general aviation to meet their informational needs often vary significantly in capability, complexity, and cost. Products range from inexpensive dial-up weather and flight-planning services that run on low-cost personal computers to high-end, customized, integrated flight department management systems running on minicomputers. But it is not uncommon to find aviation computing products that span this broad range coexisting in the same flight department of a Fortune 500 corporation.

The reason is because (so far, anyway) there aren't any regulatory or industry standards for business or general aviation's use of computer aids (with one exception, which I'll discuss in a moment). Pilots and flight departments are free to define which kinds of tasks they want to perform with the help of computer tools. For example, operators with FAR Part 135 air taxi certificates are free to do all their flight planning "the old-fashioned way"—manually, with an E6B whiz wheel (a WWII-era circular slide rule used to calculate time, speed, and distance computations). And nowhere does it say that a corporate operator must use a computerized maintenance tracking program or a multimodular dispatching system with an airline interface. In a way, this situation is surprising, given the FAA's propensity to define standards for virtually all other aspects of aviation. So far, the aviation computing industry has been self-policing. The only area the FAA has expressed any interest in certificating is navigation databases, which must meet certain accuracy criteria. And since not all corporate flight departments are fully convinced of the value of computers yet, the recreational pilot of a rented Cessna 150 might have invested more in aviation computing programs than have some major corporate flight departments, although that's becoming an unusual exception. And although it's probably atypical, that Cessna driver might have the latest Windows-based, Jeppesen NavData-based flight planner, a tax tracker, a computerized combination logbook and maintenance-cost tracking system, a CD-ROM-based Federal Aviation Regulation/Aeronautical Information Manual (FAR/AIM) reference, an account with a World Wide Web-based weather-briefing system that would put some airlines to shame, and a desktop flight simulator that can replicate nearly every detail of his or her aircraft's performance (or that of an MD-80) just for practice when the weather's down around your ankles. How many corporate flight departments can boast the same?

What's also amazing is that there is—as yet—no "Microsoft" of the business/general aviation computing marketplace. It is still a surprisingly fragmented field, with many vendors who have been selling their products for years. The business and general aviation software marketplace is equally open to the entrepreneur who can find a niche and fill it with a worthwhile product or improve an existing one. I am constantly surprised at how often start-ups have successfully brought to market innovative new products in new market niches or have shouldered their way into what I thought was an already overcrowded corner of this market. It happens all the time—if the mix of product, capital, and marketing know-how is right. And the ability of some vendors to cling to the margins of a small market year after year is equally noteworthy.

Business aviation computing solutions

The lion's share of the products for business and general aviation have been written for use by corporate flight departments. The computing products used by business aviation fall into three basic categories: (1) scheduler/dispatchers, (2) flight-planning/weather-retrieval systems,

Figure 7.2   New general aviation aircraft, such as this prototype of the Cirrus SR20, are starting to feature advanced cockpit displays, such as the ARNAV Systems flat-panel, multifunction display. (Cirrus)

and (3) a miscellaneous category that includes a variety of mostly unique products. After discussing some business aviation concepts, we examine the scheduling/dispatching systems, then the weather and flight-planning systems, because they are used by nearly every business aircraft operator in one form or another. These categories also provide some clues to the scope of this segment of aviation to those who might be unfamiliar with it.

There are far more general aviation pilots and aircraft than business aviation pilots or aircraft. So why are there more computing products for business aviation than for general aviation? The answer is simple: In the 1970s, when corporate flight department computing was still very new, one way to succeed as an aviation computing systems developer was to have (1) an interest in aviation, (2) a knack for writing computer code (such as COBOL or FORTRAN), (3) access to some corporate mainframe or minicomputer time (usually in the wee hours of the morning), and (4) a flight department desperately in need of the automation tools that a good flight scheduling/dispatching program could provide. If you had a few well-heeled corporate clients, you were set. Your clients either viewed you as a demigod or a dweeb but, in either case, a necessary key to their survival. Until the 1970s, business flight departments were generally small, run by a close friend of the company president (they might have been buddies in WWII), and computers were something used only by university scientists or the military.

In the early days of business aviation computing, you didn't need an expensive marketing or distribution scheme to succeed—just a good word-of-mouth network and maybe a small exhibit at the annual NBAA show. Another reason programmers succeeded in business aviation was that few general aviation pilots in the 1970s had access to a computer that could do much more than play Pong.

Of course, the marketplace has changed a lot in the past few years, and the emphasis might be about to favor general aviation-oriented software vendors with large marketing budgets. It will be interesting to watch.

Dick Aarons and the Origins of "Avcomps"

I've had the privilege to work for some of the seminal individuals in the world of business aviation publishing. One of them, Richard Aarons, was the man who coined the term "avcomps" (for aviation computing systems or services) and gave the business aviation computing system industry its first notice and legitimacy. Dick is best known in business aviation, however, as the editor-in-chief of *Business & Commercial Aviation,* the leading monthly magazine of the business aviation world. He's been there forever, it seems, so his grasp of the segment is encyclopedic. Dick Aarons is also someone with an eye for significant new trends. One of them was aviation computing.

According to Aarons in a recent interview, *B/CA*'s interest in computers and software began in the early 1970s. The magazine's first article on the subject tried to determine just what flight department functions ought to be computerized. In those days, the minicomputer was making it first appearance in corporate America, so it stood to reason that some flight departments might have access to computers and could make use of them. At the same time, the first microcomputers (this was before "personal computer" was a common term) began to appear—Commodores, Texas Instruments, Ataris, and the ubiquitous Tandy/Radio Shack TRS-80 were the hot new computers then. IBM's first "PCs" wouldn't appear until 1981. Hard drives were just starting to appear, which made some of these computers useful for minor database chores.

"This article led to a second article that was never published," said Aarons. "The concept was to write, in what is now called 'pseudocode,' a flight department management program. That project led to OAS 91 and the formation of a startup company called Aviation Information Services." (AIS still exists today, under different ownership.) "OAS 91 was a true tracking, scheduling/dispatch program written in COBOL, written to run on a Data General Eclipse minicomputer. The one we used at the time was Ziff-Davis' accounting computer. [*B/CA* was once owned by the Ziff-Davis publishing company.] We bought an extra 90-megabyte drive for it, which of course in those days was the size of a washing machine! We even had a party when this drive arrived on a piano dolly in the office one day. Soon after, we contracted with an outside vendor because of security concerns; then we hired a staff and brought in a sales director from Jeppesen. Our first customer was RCA's flight department."

The original AIS management information program was loosely modeled on the CSI/CAMP system for maintenance. CAMP Systems, Inc., was one of the originators of computerized aircraft maintenance programs (hence the name) for tracking engine and airframe cycles, maintenance due lists, and inspections to comply with FAA and manufacturers' recommended maintenance programs. CAMP supplies its clients with maintenance logs, to be filled out by the flight department's chief of maintenance. The records are mailed in, processed on CAMP's mainframe computer, and reviewed by maintenance specialists who then produce an aircraft status report and recommended maintenance due lists for the client to follow. The service actually cuts down on the paperwork and time spent researching aircraft logs to ensure they're in compliance with regulations. (We take a closer look at CAMP later in this chapter.)

Figure 7.3 Business aircraft, such as the Dassault Falcon 50EX, have benefitted from the digital avionics revolution. This aircraft now offers Collins' Pro Line 4 integrated avionics system. (Collins)

AIS provided its customers with flight logs to be filled out by the crews and dispatchers. The crews would fill in all the relevant information about each trip. Then the logs were sent in to AIS, where they were entered by keypunch into the computer. The program would then produce about 20 monthly management reports, such as aircraft utilization, flightcrew regulatory compliance, fuel tax rebates, crew compensation, and internal departmental chargeback billings. Although the reports were very unsophisticated by today's standards, they were the first successful effort to define and computerize tasks that until that time had been mostly intuitive exercises.

"At about this time," said Aarons, "second-generation management was showing up in flight departments and in corporations—flight departments were beginning to be run much more professionally—in a business manner, and that sparked the need for what we developed."

A year after it began, Ziff-Davis went through a management change and decided to concentrate on publishing, and AIS was sold to a venture capital group in Denver. (At about this time, too, Ziff started *PC* magazine.)

Other software developers, many of whom were either pilots or flight department managers, were recognizing the need of this market, and soon companies such as Computing Technologies for Aviation, SeaGil

Software, Andromeda, Professional Flight Management, Aviation Analysis, and Flight Watch International were formed, thrived, and improved the art and science of providing flight department managers with a unique set of computerized management tools.

More Bang for the Buck:
Operational Optimizing

If ever there was a need for "avcomps," it's in business aviation. Like software written for other corporate departments, flight department computing systems tend to be bottom-line-oriented. This isn't surprising, since the managers of corporate flight departments have three primary missions: (1) to make sure safe, reliable company aircraft and flight crews are available 24 hours a day; (2) to be able to *prove*—every time they're asked by flinty-eyed company accountants—that using a $25 million company aircraft is a cost-effective means of executive transportation; and (3) to hire pilots who won't spill the CEO's coffee while returning the chairman to *terra firma*. Business aviation software can help their users do 1 and 2, and it's only a matter of time before they can do 3.

The leading vendors in business aviation computing have focused on the scheduling/dispatching, cost-tracking, and flight planning needs of larger corporate flight departments—those with at least one large, cabin-class turbine aircraft, half a dozen pilots, a flight dispatcher or two, and possibly a full- or part-time maintenance team—but you'll find their products in many smaller corporate hangars, too, as well as in the computers of similarly sized aircraft management firms.

In the 1960s and 1970s, when many flight departments were established or began to expand, tracking multiple aircraft, passengers, and crews was done with the use of a felt-tip pen on a whiteboard (greaseboard). But there were risks to using this seemingly simple method. Last-minute schedule changes, surprise maintenance glitches, and other operational wrinkles might leave company executives with less-than-desirable service—or, sometimes, none at all. The executives usually didn't want to give up their use of company aircraft, so the computerized flight scheduler was invented to meet the need.

In their earliest iterations, computerized scheduler/dispatchers that could electronically coordinate the comings and goings of company executives with fleets of modern business jets, turboprops, and helicopters, as well as their flight and maintenance crews, were a godsend to harried flight department chiefs who were responsible for this chore. Many of these individuals (whose titles hadn't yet been elevated to the status of "aviation department managers") often had little in the way of management or dispatching skills beyond those they developed as pilots of the

Figure 7.4 The Cirrus SR20 prototype was introduced at the 1994 EAA convention. (Cirrus)

earliest business aircraft or in the military. Indeed, most scheduling was left in the hands of executive assistants (this is still the case in many smaller companies) who rarely had any appreciation for, nor understanding of, the challenges of operating a flight department. (Before I get beat up for sounding biased, I should stress that many executive assistants and corporate travel department managers have since gone to the trouble to learn more about their flight departments and the computing systems they use and are now often considered an aviation computing system vendor's primary point of contact with a company.)

On the other hand, it sometimes required persuasion to convince experienced dispatchers to accept the new electronic tools they were being given. In one instance, flight department personnel concocted a series of excuses for not removing the department's well-worn wall planning chart, which had become a kind of last bastion against encroaching new technology, before accepting that the new software could do a better job. Even today, there are some holdouts, but today's newer crop of dispatchers, schedulers, executive assistants, and corporate travel departments have embraced the computer as an invaluable "appliance" for the flight department, and familiarity with computers and advanced scheduling software is often seen as an advantageous skill, if not a job prerequisite. At a 1993 NBAA-sponsored conference of schedulers and dispatchers I attended, a very unscientific survey revealed that of those attending, some 95 percent of their flight departments used one form or another of scheduling/dispatching software. The figure is probably even higher now.

During the leaner business years of the late-1980s and early-1990s, when many corporate flight departments came under attack by stockholders who viewed company aircraft as wasteful luxuries, business aviation computing developers focused their efforts on easing the justification workload of the flight department manager. While there isn't a software program on the market that can *always* prove that company aircraft make good business sense, some products have come very close. The holy grail of business aviation management is flight department *optimization*. Optimization means a flight department manager can prove to any skeptical CFO or stockholder that the company aircraft is being utilized to its maximum potential to hold the line on the company's air-transportation costs.

This optimization concept is part of an overall, industry-led effort to prove that without use of company aircraft, many successful corporations could not function competitively. The current banner of this program is summed up in a slogan fostered by the NBAA, "No Plane—No Gain." Many business aviation computing vendors have taken the optimization concept to heart and developed new database analysis and reporting talents to prove the value of company aircraft. I discuss those shortly.

Flight Scheduling/Dispatching Systems

Today's typical state-of-the-art scheduling/dispatching system comprises a number of integrated modules employing graphical user interfaces and operate on standard PC or Mac platforms, often in networked environments.

Typical scheduling/dispatching suites include modules for aircraft and crew scheduling; flight log generation; personnel record-keeping; aircraft maintenance tracking; inventory management; financial reporting; and accounting, general ledger, and payroll. And, depending on the nature of the department's operations, the system might include an airline interface or a charter quote module. All these modules can be purchased collectively or separately and often can be customized to suit the operator's needs. (Some companies, such as SeaGil/BART, have succeeded on the economical one-size-fits-all approach, however.)

In their role as aircraft schedulers and dispatchers, business aviation computing systems rival the capabilities of a small airline's operations and reservations systems in many respects. The best of them can integrate and coordinate the travel itineraries of hundreds of corporate passengers, by their rank within the company and their personal preferences (for diet, entertainment, seating, and favorite crewmembers, for example) with the schedule requirements of a large fleet of high-performance, maintenance-intensive aircraft, the regulatory requirements of the FAA, and the personal needs of the flight crews.

Juggling a juggernaut of jets

The following fictional but typical scenario illustrates the challenges that a business aviation scheduling/dispatching system can handle with ease:

The CEO for the Colossal Candy Company decides it is time to take five members of the company's board of directors on a tour of the company's vanilla-bean processing and distribution operations in the United States, Argentina, and Pakistan. The tour will originate at the company's hangar in White Plains, New York (HPN). The company operates two Cessna Citations and a Gulfstream G-IV. The tour flight is to depart HPN at 0900 on a Monday and return the following Monday evening, so the Citations are dispatched on Sunday afternoon to gather up the directors, who live in Burlington, Vermont (BTV), Chicago (ORD), Fort Lauderdale (FLL) and Orlando (MCO), Florida, and Phoenix (PHX), Arizona.

Everything's fine until Citation A strikes a mallard duck (a rare, but not unheard-of event) on short final to BTV, requiring some extensive repairs to the jet's right wing. Obtaining parts on a Sunday afternoon might be tricky, so Citation B, which was originally slated to be used for crew recurrency training, is dispatched to gather up the passengers. Passenger 1 from BTV is whisked back to HPN. Passenger 2, at ORD, calls to say he will be delayed, but he agrees to fly commercially to Atlanta (ATL) that evening, where he will meet Citation B after it has picked up Passengers 3 and 4.

Weather and ATC delays put the flight two hours behind schedule. Can the Colossal Candy Co. Gulfstream get Passengers 2 and 5? With a few mouse-clicks on her computer, Colossal Candy's ace flight dispatcher determines that if the scheduled maintenance on the Gulfstream's avionics is deferred from tonight until the night of its return from the tour, it can be done, provided the dispatcher can find a legal crew. The company's operations manual requires that the pilots who will fly tour the next morning must have had 24 hours off-duty prior to an international flight. Who can fly tonight's trip? Another few mouse-clicks reveals a legal crew to fly to PHX and ATL. Phone calls to each of Colossal Candy's Gulfstream pilots and a call to the company's weather and flight-planning service sets the alternate flight plan in motion, and by 2300, everyone is in HPN. Got that?

While such a situation could have been handled manually, the flight scheduling system saved dispatchers from hours of replanning and obviated extra calls to maintenance personnel and searches through the Gulfstream's maintenance records. And the tour hasn't even started yet!

Weeks before the tour started, the dispatching system reminded the dispatcher to (1) stock the galley with a particular wine and samples of the latest Colossal Candy concoctions for an inflight focus group, (2) update some of the passengers' passports, (3) contact a particular, multilingual freelance flight attendant who would also act as an interpreter on the tour, (4) order some specialized navigation charts, (5) arrange for aircraft security, hotels, and limousines for the passengers and crew at their destinations and (6) obtain overflight permissions and landing approvals for the international portions of the tour.

As the tour progresses from HPN to its various stops, the schedule is checked to make sure that the aircraft and crew are still operating within their prescribed limits. At one point, the CEO tells the flight crew he wants to take his entourage to Bangkok to visit a potential new cinnamon-processing plant before returning home. What would have been a dispatcher's nightmare in the days of greaseboard flight scheduling becomes an easy copy-and-paste task with a well-designed scheduling system. A couple of phone calls and *voila!* A new itinerary.

On-the-fly flight scheduling is only one aspect of scheduling/dispatching systems. They're also essential to flight department managers as a means of gathering information about the flight department's operations and translating it into reports that can usually be understood by corporate accountants. Typically, a scheduling/dispatching system is configured to perform the following tasks:

- Track and project individual aircraft airframe and engine cycles and provide warnings if an aircraft is about to exceed a predetermined maintenance cycle

- Create fuel purchase reports—a must for any operator and essential for operators with fuel farms and those who regularly tanker fuel (carry extra fuel to avoid the expense of buying it when traveling)

- Generate passenger summary reports that provide listings of all passengers who have flown, their itineraries, charges, and their authorizers. In many flight departments, these costs are called "chargebacks" and are billed directly to the departments that used the aircraft

- Create analyses of city pairs, which can help operators identify the need to add bases, aircraft, and personnel;

- Create aircraft operations reports that quantify when and where aircraft were flown, their associated direct costs, costs per mile, costs per hour, and so on

- Create departmental operating budget reports and, depending on the needs of the operator, other ad hoc reports

Customization and information mining

The ability to digest and analyze raw information about a flight department into standard business report formats has, for the past 10 years or so, helped to elevate the role of the flight department manager from a glorified chief pilot to that of a bona fide departmental manager in many corporations that consider business aircraft an essential tool. Business aviation computing system developers have capitalized on this recent legitimacy by exploring new ways to help flight department managers survive budget cuts, mergers, and new ownerships. This "information mining," as some refer to it, is routine operational analysis jazzed up with a new metaphor, but it's an important concept in flight department computing. Let's see how it works.

For a long time, many aviation computing products were designed to provide "canned" reports for their users. In other words, a fuel purchase report, for example, created on a SkyAces Software scheduling/dispatch system for the Colossal Candy Company's corporate flight department, would look pretty much the same as one created on the Gonzo Granola Company's SkyAces Software scheduling/dispatch system. This one-size-fits-all approach to operational reporting worked for a while, until an accountant asked the Colossal Candy Company's flight department manager to provide the report in a slightly different format with a slightly different emphasis or layout. Colossal's flight department manager could have explained to the accountant that the software could only format the reports one way—or, more likely—he or she called up the SkyAces Software Company and explained the dilemma.

SkyAces Software now had its first custom client, since its data file structure was proprietary, and even if Colossal's flight department had a software whiz on staff to extract the data, it would be more expedient to "farm out" the report to SkyAces. So, for a modest fee, SkyAces programmers could extract the necessary information from Colossal Candy Company's data and create the report in the required format. And for an additional fee, other users of SkyAces Software's scheduling/dispatch system's report generator module could obtain a copy of the report format and instructions for using it.

Some software vendors have made a name for themselves by capitalizing on their ability to customize reports to any format required by their clients. In a sense, they have become "outsource specialists" for flight departments that either don't have the manpower or the time to "tweak" their software to their own needs or those flight departments that find off-the-shelf scheduling/dispatching systems too simplistic for their purposes.

Free the data! Today's graphical computing platforms, such as Windows, OS/2, and Mac System II are reshaping business aviation software once again. Users of DOS-based scheduling/dispatch systems are quickly upgrading their software to match the capabilities of their "horizontal," or nonaviation, general-purpose software, such as word processors and spreadsheets. The new graphical programs, combined with the introduction of desktop publishing and presentation generators, have so simplified the process of managing raw data that many end users are demanding that they have complete access to their own data files and be free of the need for customized reports created by third parties.

The result of this "free the data" movement is the ability of some systems to enable the user to extract any data from any part of a user's system and produce a meaningful report. For example, instead of simply reporting that maintenance expenses on one of the aircraft had risen 9 percent the previous year, an operator could now determine that Captain Jack had been assigned to that aircraft on 79 percent of its trips, and that the aircraft's brake system required service after almost every one of those flights. Either Captain Jack was fond of riding the brakes (which could be easily ascertained by Check Pilot Charlie) or the brakes were used properly for the conditions but couldn't handle the chore. Where had Captain Jack been flying on those trips? A query of the flight log database revealed that the jet had started flying into a particularly short one-runway airport in the Caribbean just prior to the spike in maintenance costs. Verdict: Alert the authorizers of the Caribbean trips that their choice of destination was costing the company more than usual. Perhaps using another airplane with shorter runway requirements might alleviate the situation. And the sharp flight department manager would seize the opportunity to show that adding another aircraft to the company fleet would not only help hold the line on costs, it could provide management with even greater transportation flexibility. *That's* information mining!

You *can* get there from here. Some scheduling/dispatch systems now include a graphical radius search feature that can be especially helpful when you know the location you want to visit but don't know which airport would be the most suitable to use. For example, the vice president in charge of Gonzo Granola's production facilities wants to visit a potential new site for a blueberry dehydration factory in Skowhegan, Maine. There is no airport in Skowhegan, but when you enter the city name, a radius of 25 miles, and a minimum runway length of (for our example) 5000 feet, you instantly see that Waterville–La Fleur airport is less than 20 miles to the south, has a 5500-foot paved runway, and provides all the facilities you might need. The search also reveals the name of a limousine service that can provide transportation for the passengers so they can avoid driving in unfamiliar territory, as well as some local eateries, lodgings, and points of interest.

Maintenance management systems

When major maintenance is required on a business aircraft, such as interior renovations, repainting, and engine overhaul work, most major corporate flight departments elect to send their aircraft to specialty centers. But most routine servicing is done in the operator's own hangar (or that of an FBO) at the home airport. While the operator is free to do the required maintenance record-keeping, most elect to do it with the help of a maintenance management system. And although these systems can be as simple as a homemade database on a small PC, most operators use a specialized maintenance management system. The company with the undisputed lion's share of the market for this kind of service in the corporate aviation world is CAMP Systems, Inc., of Ronkonoma, New York.

CSI/CAMP has been providing corporate and government operators with innovative computerized maintenance record-keeping services since the early 1970s. More than 2000 aircraft are enrolled in the company's computerized aircraft maintenance program and other CSI programs.

Each month, CSI's large staff of maintenance analysis specialists help keep their assigned clients' aircraft maintenance records up to date. The process is simple. Upon enrollment, CSI technicians work with subscriber clients to create a completely current set of maintenance records for the client's aircraft, including all of its components and systems. At the same time, either a manufacturer-recommended or a customized maintenance program for the aircraft is created. Each month, operators receive work compliance forms to fill out. As they complete tasks listed on monthly maintenance due reports, they submit the forms to CAMP, where they are processed by computer to update the history of each airplane in the program. Monthly aircraft status reports are issued, detailing the aircraft's overall status, airframe and powerplant cycle times, replacement costs, and approximate man-hours to complete each item. A maintenance due list, derived from the aircraft status report, is also issued. This list notes all the inspections, servicing, and component replacement that will come due within the next two months. Customers also receive performance, reliability, aircraft history, and yearly budget reports and other useful information. When necessary, CAMP provides its clients with lists of tools, part numbers, lubricants, torque settings, and even illustrated guidelines for each task.

When questions arise, maintenance specialists, who have ready access to one of the world's most complete aviation maintenance libraries, can quickly answer questions posed by corporate maintenance chiefs around the world. When an aircraft is sold, sellers and buyers can quickly determine a fair price for the aircraft based on its maintenance history. Descriptions of business aircraft for sale often include "CAMP," which lets potential buyers know the maintenance record is likely to be flawless—a definite plus.

The beauty of the system is that the subscriber need not even be computer-literate to take advantage of the service. CAMP uses a batch-type data-entry system that might at first seem as outdated as disco in today's world of online computing, but in CAMP's case, the system is ideal.

Some CAMP clients prefer to maintain a closer contact with their aircraft's maintenance records. These clients can subscribe to CAMP's CSI Link program, which connects them, via password-protected computer terminal, to the database at CAMP's headquarters. This way, subscribers can instantly access, add, delete, or change information in the aircraft's records. They can also obtain ADs (airworthiness directives), which carry the power of FAA regulations, and manufacturer-issued service bulletins. The system is also useful for making "what-if" scenarios for planning purposes. CSI Link users are free to mail in completed work compliance forms like any other subscriber.

CAMP has recently been integrating its maintenance products with its Andromeda scheduling/dispatching system, so customers can integrate their maintenance data with their operational planning. For example, with a maintenance tracking module as part of the operator's Andromeda flight scheduling system, a dispatcher can quickly coordinate the flight and maintenance schedules of a large, multitype fleet of aircraft.

I need to emphasize that while CAMP has captured a large majority of the business aviation fleet, other software vendors are penetrating this market with similarly capable systems. Some vendors—including CAMP—have been forming strategic alliances with business airframers to provide integrated support for the airframers' customers; CAMP has recently developed programs for operators of specific models of Learjet and Raytheon (formerly Beechcraft) aircraft. Other providers of maintenance modules or stand-alone maintenance programs include Aviation Analysis, CTA, CTV, Flight Watch International, PRG Aviation Systems, PFM, Professional Software Associates, SeaGil/BART, Silicon Wings, and SEA (through Universal Weather).

Weather-Briefing and Flight-Planning Systems

The often impromptu nature of corporate flight department operations make them among the most demanding customers for weather briefings and flight-planning and -handling services—perhaps even more so than air carriers. And since general aviation and business aviation have many common needs for weather and flight-planning services, I discuss the two segments together.

I've put the two categories of weather briefing and flight-planning systems together because their definitions usually overlap, and many vendors offer comprehensive soup-to-nuts weather *and* flight-planning services under one roof. But to make these categories more understandable, I define them individually.

Weather-briefing systems

Although all pilots must obtain comprehensive weather briefings and up-to-date information about their route of flight, destination, and alternate airports, the corporate pilot is most apt to need to perform these tasks with little notice, and indeed, might have to do it on the fly (literally or figuratively), often without the benefit of a company dispatcher. The following information about weather-briefing systems is applicable to all categories of flight operations, not just business and general aviation.

Aviation weather-briefing services available today are changing quickly. The days of the telex machine clattering away in the corner of a FBO or FSS are gone, and I doubt that few pilots miss them. Today, corporate pilots have an arsenal of mostly computer-based resources at their disposal to provide them with near-real-time weather information. Essentially, there are four main sources of flight-briefing information, each requiring varying levels of hardware or computer skills to access:

Flight Service Stations. Flight Service Stations (FSSs) in the United States and similar government-sponsored or -operated services elsewhere, weather-by-fax services, dial-up services, and online resources. Generally speaking, all one needs to obtain a briefing from FSS is a pencil and paper and telephone (unless you happen to be at one of the dwindling number of places that still have a walk-in FSS, in which case you don't even need a telephone). Most of the charts produced by the National Weather Service (NWS) for aviation purposes are usually available for walk-ins to ponder, and you can usually talk to a briefing specialist. For workload and ease of access, an automated telephone system connects callers to a briefing specialist, and, although she or he might be located far from your route of flight, the network she or he uses to pull together a weather briefing for pilots provides basically the same information whether the briefer is in Bangor or Miami. Many pilots start their weather briefings by calling FSS and augment that information with some of the others described below. Some pilots only use FSS to file or close out FAA flight plans. One major FSS plus is that one can usually be contacted by VHF radio nearly anywhere.

Weather-by-fax services. Weather-by-fax services have become a popular resource in the past five years or so, although some of the older private weather vendors have offered it for much longer (if you could afford a facsimile machine before they became so popular). Obviously, to use this service you need a fax machine—either a stand-alone desktop model, a fax/modem in or connected to a computer, or a fax-capable mobile telephone.

The big appeal of faxed weather is that in addition to textual briefing information, pilots can obtain printouts of graphical products from a large menu of essential to obscure weather products that can make visualizing the weather far easier than text-only briefings or one received over the phone from an FSS. Companies that provide fax weather products either produce these charts and maps themselves or obtain them from "weather wholesalers" such as Kavouras, WSI, or Alden Electronics. Vendors might require that users of their services also sign up for other services, such as flight planning or international flight handling. Faxes are handy because the pilot can take it to the cockpit for later review or comparison with onboard weather radar, for example.

Dial-up services. The rapid development of PC technology, quickly falling hardware costs, and, more importantly, the introduction of better, easier-to-use software has spurred the development of dial-up weather briefing services. Today, next to the Internet services we talk about in Chapter 3, dial-up weather services are the fastest growing preflight planning resources for pilots and dispatchers, whether they are business, general aviation, or other types of operators.

The first of these services appeared about 15 to 20 years ago as add-ons to services the major commercial weather vendors provided to their customers. Originally, users could only access these services by means of dedicated, password-protected, usually leased and costly computer terminals or "briefing kiosks" located in the tonier business aviation FBOs (some of these kiosks are still available). Later, companies such as Flight Data, Inc., became successful by combining a number of computerized products available from several weather vendors in one menu-driven package. For varying fees (which seem exorbitant by today's standards) users could obtain text briefings from one vendor, four-color weather graphics from another, and weather maps from yet another. The first of Flight Data's weather kiosks were only available at selected FBOs, but Flight Data now offers this service as a Windows-based product for use on any PC with a modem.

One of the biggest breakthroughs in dial-up services came with the introduction of the FAA-sponsored DUAT (Direct User Access Terminal) system in February 1990. This program, sometimes called "the glass teletype" by pilots, has enabled any pilot who has access to a computer terminal and a modem to connect to either of the two private vendors (Data Transformation Corporation and GTE DUATS) that received the FAA nod to provide the service. (It was originally envisioned to be provided by three vendors but only two made the final cut.)

DUAT service provides its users with menus of preflight weather and other briefing products that include all the familiar (to pilots) reports and forecasts, SIGMETs, AIRMETs, Center weather advisories, NOTAMs and other textual information, as well as online FAA Flight Plan filing—for free. Graphical aviation weather products from third-party vendors and other services (including airline schedules and reservations) are available for modest, per-use fees (via credit card) to DUAT users who have computers that support these types of products.

The system was an immediate hit with many pilots. In the first year, more than 2 million briefings were provided to users of the two FAA-sponsored vendors. That figure is now hovering around 3 million a year, and some 150,000 pilots have tried DUAT at least once, according to the vendors. Pilots praise DUAT because of its convenience and because it offers an alternate to Flight Service briefings. And pilots appreciate getting all the information they need to make an informed decision about a flight without the "editing" that some pilots believe they get from FSS briefers. A DUAT printout provides the pilot with an opportunity to study the information more carefully than might be possible with a phone call to Flight Service, and they can take the printout with them into the cockpit. Some general aviation pilots like the option of getting DUAT's plain-English translations of the FAA's seemingly arcane textual weather code, which dates back to pre-WWII days.

DUAT has occasionally been put up for adoption by the private sector when FAA budget constraints have called for massive cutbacks, but the users have been very vocal in supporting it as a government-provided service. For now, it will stay that way.

Most private weather providers now offer a connection to DUAT as part of their products and services. As we see in Chapter 10, DUAT-like informational displays are being tested for in-cockpit viewing. One company, Saysoft, is even developing a system to receive DUAT images via portable telephone connected to an in-cockpit, color laptop computer displaying aeronautical charts. The system will enable pilots of aircraft without radar to receive a kind of "virtual radar" by overlaying the DUAT radar graphics onto the navigational charts. A number of technical hurdles needed to be overcome when I last checked with Saysoft about this system, but it looks promising.

Meanwhile, private weather companies have continued to improve their offerings for clients who need a weather-briefing service that can be tailored to their operational needs, and they offer more products than DUAT has to offer. Further, DUAT provides service only within the United States, so operators in other parts of the world need to turn to one of the private providers, which are, generally speaking, U.S.-based. Many of these vendors offer stand-alone or multipurpose programs that automate the dial-up and menu-selection process; some vendors offer additional software modules that can be used to plan the route of flight, while others can perform this task while online. And some of the vendors provide business and general aviation operators with soup-to-nuts flight-handling services such as

- ARO (Airport Reservation Office) landing-slot reservations for high-density U.S. airports;
- International route planning
- Overflight permissions
- Catering, restaurants, and overnight lodgings
- Security analysis and provision services (i.e., hiring someone to guard the aircraft when not in use)
- Transportation
- Emergency medical evacuations
- Consular contacts

Although they don't require the sophistication of some business aviation flight-planning systems, most of the low-cost flight-planning systems sold to general aviation pilots today are nevertheless remarkably capable. In 1989, only EMI Aerocorp (available on CompuServe), and

MentorPlus and RMS Technology (each available as stand-alone packages for desktop PC use) were seriously competing for a major share of the general aviation market for flight planning. These DOS-based systems were reasonably comprehensive. Their immediate popularity has encouraged their developers to continue to improve their products and also has led to a number of equally capable and innovative imitations. The development of faster, more powerful computers, larger navigation databases, and GUI-based software (Windows and Apple/Mac, specifically) has enabled software engineers to create economical systems that today rival the capabilities of flight planners offered by mainstream business aviation software vendors. Many of these programs now feature, in addition to basic route plotting, automatic downloading of weather data from DUAT, the display of digitized aeronautical charts with overlays of the flight plan, aircraft weight-and-balance calculation (and warnings if an out-of-limits condition exists), extensive performance databases for a wide variety of aircraft (ranging from light singles up to large business jets), in-cockpit laptop computer portability, and, via a simple connection to an onboard GPS (Global Positioning System) sensor (via an RS 232 connector), a moving-map display. These kinds of capabilities were the stuff of dreams in 1985.

Online resources.   The last category of weather-briefing resources is online computerized networks and related systems. With the growing popularity of the Internet and World Wide Web, business and general aviation users are finding access to a host of new, exciting, and economical weather and flight-planning resources. These resources are described in detail in Chapter 3.

Computerized flight planning

Until as recently as 1985, there were only two ways to plan a flight. The first method was to do it manually. A pilot or dispatcher could sit down with an E6B whiz wheel, a pile of charts, the performance tables for the aircraft, and a few pencils and manually calculate the aircraft's headings, speeds, altitudes, and times; determine which navaids and radio frequencies to use; determine the minimum altitudes for terrain clearance; calculate how much runway was required for takeoff and landing; how much fuel was required; how much payload (passengers and cargo, if any) could be carried and where to locate it; find a suitable alternate airport if the primary destination's weather was in question, and so forth, until all the blank spaces in a flight log were filled in. Then he or she hoped conditions didn't change as the flight progressed. For updates, you could call FSS or your dispatcher on the flight phone.

If you went to the same airports all the time, this manual flight-planning exercise usually wouldn't be too demanding, but it could be time-consuming. But if you were a busy flight department (particularly one that operated internationally), you turned to an outside flight-planning service provider, as a number of airlines did before they entered the computer age in the 1970s.

An account with a weather and flight-planning service, such as Lockheed DataPlan, Universal Weather, Global Wulfsberg, or EMI, enables you to store your aircraft's performance data on its computer. Once enrolled as a client, all that you need to do to obtain a flight plan and a flight log is to call up and tell the briefer (who is most likely a licensed aircraft dispatcher with a background in meteorology) where and when you want to go, and he or she does the rest. The vendor's computer optimizes your flight performance, depending on wind, temperature, runway, field elevation conditions, and whether you wanted to fly to destination at maximum speed or with a maximum load of fuel or passengers. You can also select the type of routing—via airways, great circle, direct, recommended routings, or a combination. In a short time your flight log is sent to you via telex or ARINC or even by fax, or it appears like magic on a dedicated flight-planning computer in your FBO or hangar. Some services even provide ARINC, SITA, AFIS, etc. telecommunications links. Today, you can communicate with many of these vendors via fax and e-mail.

Another advantage to using an outside service bureau are the additional services they offer. Most of the larger firms can provide their clients with international trip-handling services that include obtaining cross-border overflight permissions, international flight-plan filing, customs advisories, security analyses, professional hand-holding for first-time international operators, and agent services. Agents are individuals hired or contracted by the handling company to be its representative at many airports around the world. An agent is similar to a hotel concierge—she or he is usually fluent in local dialects and customs, capable of smoothing over many potentially sticky situations, and helpful in preventing confusion when it comes to fueling, finding on-the-road maintenance facilities, making hotel and transportation arrangements, providing translation and monetary exchange services, dealing with local aviation authorities, and making recommendations about where to obtain safe, if not always savory, food and water for the passengers and crew.

Today a variety of flight-planning options are available to business and general aviation operators. You can still create a flight plan manually, of course, or contract with a flight planning/handling service to provide some or all of its services, depending on the requirements of your trip; or you can use a stand-alone software package that operates on a PC.

Examples of today's weather and flight-planning systems. The following companies are among the outstanding providers of today's computer-based weather and flight-planning services used in business and general aviation:

- Accu-Weather, Inc., provides worldwide weather briefings, international NOTAMs, satellite and NEXRAD radar imagery, DUAT graphics, and fax weather.

- Alden Electronics furnishes its weather system users with a variety of products, such as lightning strike data (showing the location, amplitude, polarity, and time of origin from a national lightning detection network), satellite weather imagery, Doppler mosaic, and NEXRAD radar imagery.

- AlliedSignal General Aviation Avionics provides a service very popular with (and tailored for) business aircraft operators, called AFIS (Airborne Flight Information System). This system is an airborne and ground-based flight information service that interfaces with some of the company's more popular aircraft flight and navigation management systems, as well as those made by other avionics manufacturers. Prior to flight, AFIS provides flight planning, FAA flight plan filing, AROs, predeparture clearances, textual and graphical weather, and international handling services. Preflight access to AlliedSignal's Global Data Center (GDC) computers requires the use of a PC. The user is connected to GDC's computer via Tymnet or CompuServe networks and inflight via ARINC, Air Canada, SITA, and AVICOM VHF networks and the Inmarsat satellite network. En route flight plans, weather, and performance updates can be displayed on the aircraft's navigation displays. Text weather and messages can be copied to a cockpit or cabin printer or terminal. The system also provides down- and up-calling capability, as well as a variety of commercial flight-planning and support services. AlliedSignal provides a range of subscription and pay-as-you-fly plans for AFIS.

- Baseops' AirPlan, originally developed for the European air charter community, features worldwide navigation planning, including NAT (North Atlantic Tracks) and ETOPS (extended twin-engine overwater operations), multiuser access, graphical displays of winds, global navigation updates, AFIS/SITA fax interface, and other options. The system's NAT module is updated daily and specifically designed to take into account the many different rules for North Atlantic operations. Once a flight plan is generated, an ICAO or FAA flight plan is sent automatically to the appropriate agency.

- EMI Aerocorp, Inc., available on the CompuServe Information System, provides easy-to-use flight planning and weather briefings, FAA flight plan filings, and FBO and hotel databases.

- Excel Software's TAU (The Aviator's Utilities) is a desktop computer-based graphical flight planner and provides access to DUAT and Accu-Weather.

- Flight Data, Inc.'s DUAT/Plus for Windows offers automated and scripted DUAT weather briefings and FAA flight plan filing. Flight Planner for Windows, another dial-up system, connects users to a variety of private online weather and flight-planning services.

- Jeppesen Sanderson's JeppFax service products include the company's aviation weather graphics, satellite imagery, high- and low-level significant weather, wind and temperature forecasts, tropical storm products, and NWS graphical and text products, among many others. Jepp's MetPlan and JetPlan services provide access to Jeppesen meteorologists 24 hours a day for personalized briefings. Jepp's international flight-planning service is available 24 hours a day for one-call flight planning and handling services.

- MentorPlus Software's FliteStar is a PC- and Mac-based graphical flight planner with automated dial-up to DUAT and Jeppesen's weather-briefing services, moving-map displays, and other useful features.

- RMS Technology's Flightsoft is a PC-based graphical flight planner with automated DUAT dial-up capability, moving-map displays and other helpful features.

- Universal Weather and Aviation, Inc.'s flight-planning and -handling services are available on the Internet at the company's World Wide Web site, but their Pilot's Choice/WindStar dial-up software for flight planning is still available for those who prefer to do their flight planning on their own computers.

- WSI's PILOTBrief satellite weather-briefing system delivers radar imagery and Jeppesen DataPlan custom aviation charts, and a modem will connect you to WSI's database of custom weather products. WEATHER For Windows, a dial-up system, provides automated connection and downloads from WSI's worldwide weather system and other features.

Miscellaneous computing systems

The business and general aviation segments have created numerous (often temporary) niche markets for savvy programmers who understood the needs of the market. The following are some of the more enduring or unusual computer systems. A complete list is included in Appendix A.

- Ac-U-Kwik provides subscriptions to comprehensive databases of worldwide preferred and direct navigation routings, FBO listings, hotels, ground transportation providers, and so on.

Main Flight Entry Screen

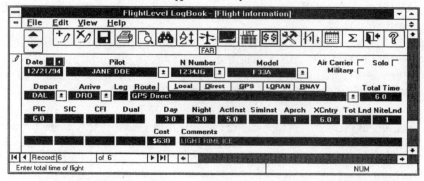

Currency Screen

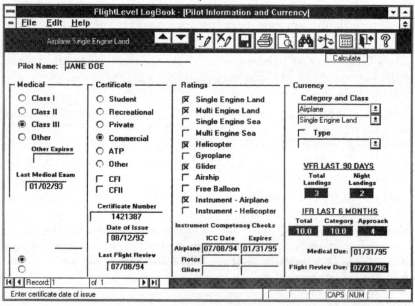

Figure 7.5 Paper logbooks are giving way to graphical computer versions that can ease the chore of tallying up pilot flight times. (Flight Level Corp.)

- Aeroprice Software's Aircraft Appraisal Software calculates general aviation aircraft wholesale and retail values.

- AIC's Performance Plus! makes operating cost comparisons for general aviation aircraft. Prospect! provides sales management tools for aircraft brokers and dealers.

- Aviation Software and Analysis's Computer Optimized Tankering System provides fuel tankering management.

- AVTAX Aviation Systems provides IRS case-tested useage determination for tax deduction purposes.

- Conklin & de Decker's Access system is the computerized version of C & D's Aircraft Cost Evaluator operating cost handbooks for business aircraft. The Life Cycle Cost Analyzer determines long-term ownership costs and financial analyses of business aircraft.

- Official Airline Guide's OAS Flight Desk provides business travel departments with information on direct and connecting flight schedules and other useful travel information. It is updated monthly.

- Flight Level Corporation's Logbook is a Windows-based database program for computerizing the drudgery of keeping pilot logbook records complete. It also provides summaries for numerous types of aircraft and operational classifications.

Air Traffic Management Systems: At a Turning Point?

The U.S. air traffic control system is in the midst of a major evolutionary transition—possibly the greatest one since its inception. The FAA has recently been under attack from all sides for not adequately providing the service for which it was conceived. The agency has, at times, been surprisingly candid in its self-criticism. At other times it has been pointedly contentious with its critics in the press and on Capitol Hill.

What follows is a look at an agency in transformation. Created in a precomputer era, the FAA is trying to catch up with a revolution in information processing. Only time will tell what the next ATC configuration will be, but it is nearly certain it won't resemble its current form. And the capabilities of the computing systems the agency is—and will be—using are often at the center of the ATC debate.

This chapter looks primarily at the state of affairs of the U.S. ATC system. While other civil aviation agencies around the world are grappling with many of the same issues, the FAA, because of its sheer size, complexity, and budget, is at the forefront of most of the major ATC issues of the day. The rest of the world's civil aviation agencies and global aviation organizations will doubtlessly model much of their future ATC infrastructures based on what works and doesn't work in the United States.

This section is neither an indictment nor an endorsement of the FAA or any of its employees. I've culled together materials from a variety of sources to present an overview of the current state of affairs at an agency that probably processes more aviation-related information than any other organization in the world. The fact that it does it as well as it does is testimony to the dedicated efforts of thousands of people in the agency. The fact that it is often seemingly overwhelmed by that effort is an indication that the agency's information management abilities are overdue for an overhaul.

Castles in the Sky

Sometime early in aviation history, the problem of how to coordinate the flight paths of two or more aircraft flying close to each other began to attract the attention of pilots and aircraft operators. Initially, the need for aircraft separation had been nonexistent because there were so few aircraft, and pilots only flew in conditions where they could see and be seen by others (this was called contact flying). For the first two decades of this century, the concept of "airspace" was more of a legal notion than an operational problem. Following World War I, however, aircraft took to the skies in rapidly growing numbers. The development of air commerce and airmail delivery led to creation of a rudimentary "bonfire and beacon" airway system that lasted until the outbreak of World War II.

The development of "blind" flying techniques in the 1930s extended the utility of aircraft and required development of new technologies to enable pilots to find their way in all-weather conditions. Low-frequency four-course radio ranges, introduced in the late 1920s and early 1930s, were the first successful method to guide an aircraft using radio signals. This method consequently required development of more organized traffic coordination methods, including the use of flagmen, light signals, and rudimentary radio communications. The first control towers, constructed to serve the needs of the proto-airlines, appeared at a number of airports in the 1930s. The traffic coordinators did little more than monitor local air traffic and advise pilots of other airplanes in the area. The concept of "controlling" aircraft from the ground would have been laughable, considering the level of technology at the time.

Meanwhile, air travel started to catch on with the public, and airlines were formed and flourished. To handle the increasing numbers of aircraft, an informal method of coordination among the airlines flying along heavily traveled air corridors (between Washington, D.C., New York City, Albany, Boston, and Chicago, for example) evolved into the first air traffic "centers." Private aviation (now called general aviation) was considered merely a nuisance to the airlines (a lot of folks would say not much has changed over the years!) Politics, airline competition, and a lack of consensus about what constituted fair and efficient management of airspace prevented development of a coherent system during this period.

It took a rash of dramatic aviation accidents in the mid-1930s before the federal government was spurred to action, following the passage of the Civil Aeronautics Act of 1938. The law created the Civil Aeronautics Authority (CAA), which was a part of the Department of Commerce. The CAA had the authority to grant airline routes and regulate airline fares. This role would soon evolve into the independent Civil Aeronautics Board (CAB), which lasted until airline regulation ended in the 1970s. The CAA later became known as the Civil Aviation Agency. (Generically speaking, all governmental civil aviation agencies are called CAAs.)

Controllers tracked flights manually on large blackboards and, later, with small strips of paper in holders made for the purpose. Today's flight strips, which are now computer-generated, provide controllers with information about the identity, destination, and routing of each flight in progress; they are the descendants of this original system. Flights were separated by predetermined times, and pilots were expected to arrive over airway fixes within a small window of time. Considering the level of meteorological technology at the time, it is remarkable how well this system worked, and, indeed, much of today's ATC structure is still time/fix based.

The U.S. CAA, however, was responsible for certification of aircraft and pilots, safety enforcement, and air traffic control. The CAA lasted until 1958, when it became the Federal Aviation Administration. This action took the role of safety rule-making away from the CAB, and accident investigation came under the auspices of the National Transportation Safety Board, where it still resides.

Nevertheless, it wasn't until the introduction of radar in the late 1940s before air traffic controllers began to achieve a real-time grasp of what was actually taking place in the sky. Equipped with this new tool, air traffic managers began to concoct a variety of schemes for ensuring the safe, efficient use of airspace. In the late 1950s, large, slow (by today's standards), vacuum tube-based computer systems first appeared, mostly as statistical tools and simple communication links. Still, it wasn't until the mid-1970s before transistor-chip-based computer technology became a useful tool for separating traffic. When these computers shut down, however, controllers still resorted to manual techniques that were developed in World War II.

The ATC system has continued to evolve somewhat, and some new technologies and traffic management techniques have continued to ensure that the U.S. ATC system is the best in the world. But many believe recent advances in communications and navigation technology have made it possible to rethink how we manage air traffic. It's time now to look at today's ATC system.

The Current U.S. ATC System

The primary function of the U.S. ATC system is to maintain safe separation of aircraft flying in the National Airspace System (NAS). Second, it is ATC's task to keep aircraft traffic moving as efficiently as possible throughout the system. In short, ATC is aviation's traffic cop, working to ensure that aircraft do not run into each other and that traffic moves in an orderly fashion with a minimum of delay.

There are several types of ATC facilities around the United States. These facilities include airport control towers, terminal radar approach control facilities (TRACONs), air route traffic control centers (ARTCCs—sometimes referred to as "centers"), the ATC Command Center (often referred to as "Flow Control"), and flight service stations (FSSs).

Networks

The world's ATC systems are tied together with a number of telecommunications circuits. The FAA's primary circuit is the National Aviation Data Interchange Network (NADIN). NADIN has two hubs, one in Salt Lake City, Utah, and the other in Atlanta, Georgia.

NADIN, in turn, is connected to a global network called the Aeronautical Fixed Telecommunications Network (AFTN). The AFTN is used to forward FAA/CAA flight plans, aviation weather, notices to airmen (NOTAMs) and other operational information. That system operates at varying speeds (at a blistering 75 to 2400 baud!), depending on which part of the world you're in, the state of the technology being used (primarily 1960s-era teletype, although X.25-category circuits are becoming more prevalent), and the kinds of data being carried over it. The U.S. National Weather Service and other global weather agencies also share space on the AFTN, for relaying weather information from one part of the world to others, although much of that data transfer function eventually will be transmitted by a recently implemented satellite network.

Facilities

Airport control towers are responsible for aircraft while they taxi to and from runways, during takeoffs and landings, and when they are in the airspace within 5 miles laterally and up to 3000 feet above the airport. The FAA bases its decision to build and operate a tower on the number and type of aircraft operations at a given airport. More than 400 U.S. airports currently have such towers, and the FAA operates most of these (a growing number are run under contract to the FAA). An FAA program called Tower Control Computer Complex (TCCC) is expected to begin implementation at a limited number of towers by 1997. TCCC will automate and integrate a number of tower-related functions.

TRACONs are responsible for controlling aircraft shortly after takeoff and prior to landing, or during the climb and descent phases of flight. There are 184 TRACONs, less than the number of towers because some TRACONS handle more than one airport. For example, a single TRACON handles the traffic approaching and departing from all three major New York-area airports as well as the other satellite airports in the New York Metroplex. TRACON controllers are assigned different sectors to manage, and each sector coordinates with adjacent sectors, as well as the airspace controlled by the ARTCCs.

The 21 ARTCCs cover even broader areas. Their job is to keep track of aircraft while they are "en route," or during the high-altitude, cruise phase of their flights. "Centers" are located in Albuquerque, Anchorage, Atlanta, Boston, Chicago (the busiest center), Cleveland, Denver, Fort Worth, Houston, Indianapolis, Jacksonville, Kansas City, Los Angeles, Memphis, Miami, Minneapolis, New York, Oakland, Salt Lake City, Seattle, and Washington, D.C.[1] Oceanic Control Centers for the Atlantic and Pacific Ocean are colocated in New York and Oakland Centers.

Basically, three computer systems are used by centers—host computers, Display Channel Complex (DCC), and the Direct Access Radar Channel (DARC). The systems are integrated for redundancy, so that if one fails, the others can continue to work, albeit at reduced capability.

The host computers, IBM 3038 mainframes installed in the mid-1980s, are the nerve center of the operation. They handle all flight-data processing, including flight plans, automated handoffs, and generation of flight progress strips used by controllers to keep track of aircraft in their assigned airspace sectors.

From the host computers, data is fed to the DCC. The aging DCC systems (IBM 9020E minicomputers) process radar information and drive the controllers' displays. The FAA will be replacing the oldest of these systems at five of the nation's busiest centers, under a $10 million contract with Loral Federal Systems, by late 1997.

Many of the nation's DCC computers will be enhanced by the interim Display Channel Complex Rehost (DCCR) and will eventually be replaced by the Display System Replacement (DSR) program. The DCCR essentially takes old software from the vintage IBM 9020E computers and puts it on IBM 9121 computers, which should give the agency at least another decade's use of this system, if needed. (Some observers equate this move to putting DOS version 2.1 software on a Pentium PC.)

A prototype DSR system will be ready for the field in March 1997, and operational systems will go into Chicago Center in October 1997, followed at one-month intervals by installations at Dallas/Ft. Worth, Washington, and Cleveland. The last system is slated to be installed in New York Center in February 1998. The DSR is projected to go operational at these ARTCCs at one-month intervals beginning in February 1999.

Figure 8.1   The Display System Replacement (DSR) will provide air traffic controllers with new advanced workstations at 20 ARTCCs. Interest in new air traffic management methods, such as Free Flight and the National Route Program, might call for use of computer-based traffic "conflict probes" and other computer aids. (Loral)

The DSR will use RISC (reduced instruction set computing) flat-panel, high-resolution display consoles. The systems will also be able to display next-generation weather radar (NEXRAD) data on the controllers' 20-inch situation display screens. Software for DSR consoles is written primarily in Ada code and comprises some 428,000 lines of source code. The controller workstations will be fitted with trackballs and or keyboards.

However, operators of U.S. airlines, as well as the National Air Traffic Controllers Association (NATCA), have expressed concerns about the reliability of the nation's ARTCCs until the DCCs are replaced and recommended that the FAA rehire retired technicians until the new equipment is operational, and ensure that parts are available at each site. The Air Transport Association said the FAA should consider contracting to fabricate new spare circuit boards, and FAA aircraft should be used to send personnel and parts needed to a site experiencing trouble.[2]

The final component in the triad is the DARC. DARC is the backup system that sends comparatively "raw" radar data directly to controller workstations. If the DCC fails, the system defaults to DARC/host operations. If the DCC and the host computer both fail, the system goes to DARC-only mode, the most degraded condition short of total shutdown.[3]

Another key facility, overseeing the entire ATC system, is the FAA's Air Traffic Control System Command Center (ATCSCC), also known as "Central Flow Control," located in Herndon, Virginia.

Flow Control is a 24-hour, seven-days-a-week facility responsible for real-time command, control, and oversight of air traffic activity in the U.S. NAS. Its role is to anticipate situations that will create bottlenecks or other problems in the system, then respond with a management plan for traffic into and out of the troubled sector. For example, if bad weather develops or a runway is closed for repairs, Flow Control will manage the number of aircraft operations into and out of the affected area. The objective is to keep traffic levels in the trouble spots manageable for the controllers.

Figure 8.2   FAA's Flow Control facility in Herndon, Virginia, provides ATC system managers a command post that was lacking for years. (FAA)

The facility was established in 1970 to regulate the flow of air traffic nationwide in order to maximize system efficiency. Today, the Command Center manages airspace for some 150,000 daily flight operations. Flow Control works in partnership with the airline industry to plan more fuel-efficient, time-saving flight routes. This partnership, says the FAA, benefits air travelers with reduced delays. Traffic-management decisions are based on computerized flight data reflecting the most accurate information and weather conditions. Flow Control traffic managers initiate ground delays at departure airports, as well as "in-trail" spacing to avoid the use of holding patterns, which are considered the least efficient method to "stop" traffic.

Prior to 1987, controllers in the 21 centers had little direct knowledge of traffic in adjacent centers that might be headed their way. To address the problem, the FAA developed the Aircraft Situation Display (ASD) to provide air traffic managers with a real-time overview of all IFR traffic in the nation's airspace. The system was made available to air carriers in 1991, and recently it is being offered to corporate aviation flight departments and other users. The ASD tracks every instrument flight from every tower and ARTCC. Based on weather and computer data, the ASD provides controllers and traffic managers with near real-time images of domestic and international commercial, private, and military flights—more than 4000 at a time are not uncommon. ASD data is also sent to airlines and other private organizations so they can monitor the system in near real-time.

The Command Center is home to other components. These include

- The Central Reservation Facility to coordinate special airspace uses (such as military airspace and practice routes)

- The Airport Reservation Office (ARO) to allocate landing "slots" at designated high-density airports and airport resources during special events such as the Super Bowl and the Olympics

- The Central Flow Weather Services Unit, which provides weather expertise to flow control managers and other users

- The National Maintenance Coordination Center, which monitors air navigation and landing aids

- The Airways Sector Field office for in-house equipment service[4]

The Volpe National Transportation Center in Cambridge, Massachusetts, also plays a key role in the ATC system by relaying data from the centers to Flow Control and back. Fortunately, if that network—called ETMS (Enhanced Traffic Management System)—were to "crash" (due to natural or manmade reasons, for example), the ATC system wouldn't come to a grinding halt. Traffic managers would resort to telephone communications with Flow Control and adjacent centers, as they frequently do already for other reasons. (One controller I recently met said, "The system might even rise to new heights of productivity" if such an outage were to occur.)

Flight service stations (FSSs) are information centers for pilots flying in and out of small cities and rural areas and inflight resources for certain types of weather and operational information. In addition, flight service stations assist in emergency situations, initiating and coordinating searches for missing or overdue aircraft. The FAA's FSS consolidation effort has trimmed the number of FSSs from 318 in 1978 to 97 in 1995, with a final target of 61 "super" automated facilities.

Agency candor

By its own admission, the FAA says its communications networks are

> fragmented, with logically and physically separate networks used for operational and agency voice and data communication services. This set of networks is costly, difficult to manage, and in many cases, provides redundant services. Individual networks within the FAA's communication system use outdated technology, are costly to operate and maintain, and perform below required performance levels or below the performance levels of comparable systems available within the FAA through government leasing contracts or through commercial vendors. Near-term savings are expected when individual networks are decommissioned and replaced with services provided by another FAA owned network or leased from a commercial vendor.[5]

NAS Structure

To understand the changes taking place in ATC, as well as the opportunities in ATC computing systems, it will be helpful to take a look at the current structure of the U.S. National Airspace System (NAS).

The NAS is comprised of a variety of airspace categories, which are dependent on altitude, proximity to airports, traffic density, and other operational factors. Some airspace, such as that which surrounds the White House in Washington, D.C., and Cape Canaveral is classified as Prohibited, and no civil aircraft may operate in such airspace.

Class A airspace exists between 18,000 and 60,000 feet over the lower 48 states. Aircraft operating at these altitudes must be operating on an FAA flight plan and maintain two-way radio communications with ATC. Class B airspace (formerly known as Terminal Control Areas) exists around high-density airports and require a specific ATC clearance, minimum aircraft equipment, and pilot qualifications. Class C (formerly known as Airport Radar Service Areas) airspace is designated in areas with significant volumes of traffic and radar facilities. Class D airspace exists around airports with operating control towers. Class E airspace covers most of the rest of the continental United States, with the exception of Military Operating Areas, Warning Areas, Alert Areas, and international airspace. The FAA can also issue temporary flight restrictions in areas where there has been a natural or manmade disaster or other events that would attract a lot of aircraft (fires, earthquakes, explosions, spacecraft launches, open-air gatherings, etc.).

Oceanic airspace is divided into flight information regions (FIRs). These regions are governed by international laws and monitored by various ATC agencies. The airspace types and operating rules in most countries are similar to those in the United States.

Aircraft altitude assignments are generally based on the aircraft's heading for en route cruise—odd altitudes for eastbound flights and even altitudes for westbound flights. Below 18,000 feet, aircraft operating on instrument flight rules (IFR) operate at thousand-foot levels (e.g., 4000, 5000, 6000, etc.). Aircraft operating on visual flight rules (VFR) operate at thousand-foot-plus-500 altitudes (e.g. 3500, 4500, 5500, etc.). Additionally, vertical separation requirements dictate that aircraft on instrument flight plans be separated by 1000 feet up to 29,000 feet. Above that altitude, vertical altitude separation increases to 2000 feet in most areas. Lateral separation requirements vary, but generally, controllers try to keep aircraft a minimum of 3 to 5 miles apart.

ATC's Tool Kit

Radar is ATC's primary tool to keep track of aircraft flying over the United States. Radar transmits radiowaves of ultrahigh frequency that "bounce" back to their source when they hit something solid. The returned signal, or radar "echo," is then analyzed by the receiver to determine both the distance and direction of the object hit. The FAA requires that virtually all aircraft flying in U.S. airspace be equipped with transponders. A transponder can sense radar signals and replies by transmitting an amplified, unique radio signal directed toward the source of the signal received. The return signal not only is stronger, but it contains a discreet four-digit code that identifies the aircraft to ground radar station. A type of transponder known as Mode C, which is used aboard most aircraft, also is capable of encoding the aircraft's altitude onto the return signal. Controllers on the ground then know how far away the aircraft is, how high it is, the direction it is headed, the type of aircraft they are "looking" at, and, if applicable, the airline operating the flight.[6]

Another type of transponder, known as Mode S, is used for air-to-air and air-to-ground collision avoidance systems called TCAS (Traffic Alert Collision Avoidance System). There are two types of TCAS—

TCAS I and TCAS II. TCAS I provides warnings (traffic advisories, or TAs) to flight crews of TCAS I-equipped aircraft of the presence of other aircraft entering within a predefined volume of airspace. No avoidance advisory is provided.

TCAS II provides a warning to flight crews of TCAS II-equipped aircraft of the presence of other aircraft entering within a predefined volume of airspace (within approximately 30 miles of the aircraft). TCAS II provides traffic advisories (TA) and resolution advisories (RAs), which recommend maneuvers to fly (turn right/left, climb or descend, maintain climb or descent, or don't climb or don't descend, for example), and coordinates RAs with other TCAS-equipped aircraft to avoid "mirror" responses to RAs, plus other filtering capabilities to avoid false alarms.

FAA's Challenges

Do any of these recent headlines sound familiar?

"FAA Computer Glitch Causes Delay at Dulles"

"Controllers Guiding Aircraft with Grease Pencils Instead of Computers"

"Safety Board Calls for ATC Fixes"

"ATC Retirements Raise Concerns"

"FAA Probes ATC Outages"

For years, the news media have been sounding a lot like Chicken Little—declaring that the sky is falling—almost literally. The U.S. ATC system is falling apart, they maintain. FAA controllers are using antiquated vacuum tube-era computers that can't be patched together anymore; aircraft are in imminent danger of colliding over our heads, as they did in the 1950s, when aircraft did indeed collide over the Grand Canyon and New York City. Close calls—so-called near-misses—are commonplace and often go unreported. And the "evidence" of a systemwide failure is compelling. There have been significant power outages in recent years in Chicago, Washington, San Juan, Dallas/Fort Worth, Miami, New York, Pittsburgh, and Los Angeles. The problem is often attributed to the power conditioning system (PCS) of the affected centers, which are supposed to sense the loss of commercial power and switch to batteries but have failed to do so. The FAA has been installing new PCSs at its ARTCCs.

On the other hand, the FAA says the ATC system is the finest in the world. Operations (defined as a movement of one aircraft from one place to another in the ATC system) are up, airport capacity is rising, controller errors are declining, and system outages represent only a tiny fraction of system availability. And an article in a January 1996 issue of *Aviation Week & Space Technology* reported:

> [The] National Transportation Safety Board said yesterday that while it remains concerned about the problems FAA is experiencing with its aging computers, the air traffic control system is "very safe" and the public "should not be unduly alarmed by recent press accounts of specific" equipment malfunctions at air route ATC (ARTCCs) centers. The Board's conclusion follows a special investigation, begun last September, of ongoing computer and related equipment outages, particularly at the five ARTCCs equipped with the aging controller display computer systems.[7]

But if you're a pilot, you have likely encountered situations that made you wonder if the system is on its last legs—controllers missing calls, lost flight plans, separation from other traffic that left you wondering if the controller's scope was working, TCAS alerts, taxiways backed up with traffic, and other glitches. If you fly in Europe, the situation is not only similar and occasionally exasperating, the ATC user fees there make it incredibly expensive.

And then you'll have a flight or two—maybe even more—when everything seems to go exactly the way it's supposed to.

If you are trying to run an airline, the costs of system delays are just beginning to be understood, but they run into the billions of dollars annually.[8] The need to make the system more efficient is imperative.

If you're a passenger, such conflicting perspectives can only be disturbing. Is the system as safe as it could be? Probably, given the level of technology it is using. But how much longer it can continue is a disturbing question.

If you're in the ATC business—as a controller, technician, human factors specialist, researcher, programmer, shift supervisor, or an associate administrator, you might wonder how much longer the amalgam of programs can continue to limp along.

A message I discovered on a recent visit to one of the FAA's many World Wide Web sites expressed what some in the agency must be feeling about the agency's struggle to build a coherent vision:

> A major cause of [ATC computer] requirements volatility is this constant failure by Programs to define an architecture up front, leaving the door open to endless haggling for "one more capability." If you don't know what architecture you want, any design will get you there. Design follows capabilities and grows with about as much structure as moss on a tree. The overall "design" resembles a random collection of loosely coupled software components, each supporting only one or two capabilities. Using structured coding we can do a fairly decent job of coding this rat's nest design and rest assured that all of the bugs will be in the design.
>
> We have reason to hope that there are now people actively trying to bring architectural vision to the FAA. Go to it gang!

The FAA, by its own admission, is in need of repair:

> With regard to the NAS, the FAA faces two major challenges in the immediate future. One is internal, while the other is external.
>
> The *internal challenge* is the FAA's need to address an aging NAS infrastructure that increasingly is difficult to maintain, and to do so in an environment of economic austerity. Because of fiscal constraints, the FAA cannot afford to replace and operate the NAS in its current physical form, and it no longer can afford (in terms of cost or time) to pursue its traditional development-oriented approach to acquiring new systems and equipment for the NAS. Also, of growing importance is the FAA's need to defend vigorously its assigned portions of the radio frequency (rf) spectrum against encroachment and to find means of using its rf spectrum more efficiently.
>
> The *external challenge* is the FAA's need to accommodate a rising number of airspace users, and to do so in a way that removes operational constraints perceived by the users as antiquated in light of today's available technology and aircraft capabilities. For scheduled air carriers, the current rigid procedures and delays caused by NAS capacity limitations inflict economic hardships that could be avoided were more freedom available to users in planning and executing their flight operations. The aviation industry has coined the term "Free Flight" to describe an emerging concept for the NAS that offers the prospect of greatly increased operational flexibility for airspace users.

We discuss Free Flight later in this chapter. The FAA goes on to say:

> In the past, the term "end state" was used by the FAA to describe an envisioned NAS in which the acquisition programs of the National Airspace System Plan were completed and provided the technological basis of NAS operations for the indefinite future. It now is understood that the combination of rapid technology development, new operational concepts, international issues, increasing economic pressures, and changing business strategies render [unfeasible] any long-range plan that fails to incorporate a process for responding to dynamic and unforeseeable circumstances.
>
> The NAS architecture recognizes that continuing change, both foreseeable and unforeseeable, is certain. To capitalize on technology improvements, it recommends that provisions for regular technology upgrades be included in the life cycle plans and budgets for NAS systems and equipment. More important, however, it embraces a process for ongoing refinement of its own products. This process will assure that changes to the NAS are evolutionary rather than revolutionary (i.e., frequent small course corrections rather than infrequent large ones), thereby providing an environment of stability and predictability for planning and investment by both the FAA and the aviation industry. The term "rolling architecture" has been applied to this process.

And finally:

> *NAS Commercially Based.* The future NAS must utilize commercially-developed systems, equipment, services, and standards to the greatest

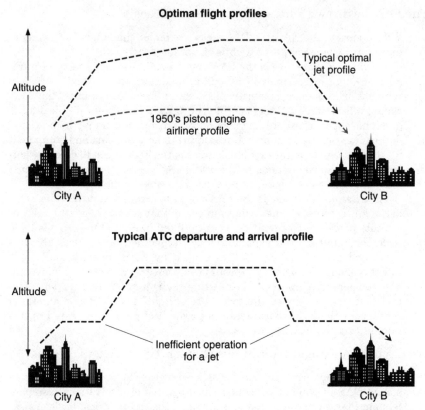

Figure 8.3   Free Flight proponents correctly point out that current "positive control" air traffic management techniques are very inefficient and based on propeller-era aircraft operations models. (RMB Associates)

extent practical. The rapid rate of technology advancement, the prohibitive costs of development programs, and the movement towards global integration of air traffic management will not permit otherwise. In addition, some user services currently provided by the Federal government may be transferred to the private sector.[9]

The ATC system, it seems, is at a crossroads. And change—big change—is in the wind.

TCAS

When I'm not writing about aviation computing systems or other subjects for various publications, I have the pleasure to serve as the senior sysop, or forum manager, on the Aviation Week Group Forum on CompuServe. This forum is an online discussion "message board" where aviation professionals can come together in a virtual space to exchange news and views about a variety of aerospace-related topics, such as training, safety, air carriers, military aviation, and ATC.

Recently I have had several exchanges with air traffic specialists on various aspects of ATC, following a report of a near-miss in the vicinity of Los Angeles International Airport (LAX). I asked one of the controllers how he'd rate TCAS (Traffic Alert Collision Avoidance System) as a safety tool.

"We know TCAS is there," he said, "and in most cases it is more of a suspect or 'unwelcome guest' than anything else. Regardless of how controllers feel about TCAS (I personally think it's a good idea when used properly), we CERTAINLY aren't going to sit back and relax under circumstances where aircraft are coming nose-to-nose!"

In the early stages of TCAS implementation into the air carrier fleets, there were some growing pains and technical snafus. One of these was the problem of "mirrored" responses when two TCAS-equipped aircraft were in close proximity. To put it simply, in a mirror situation, the software was causing both aircraft to perform the same evasive maneuvers, in a mirror image—each airplane would descend, for example, to avoid penetrating the other aircraft's protected airspace. Naturally, this situation only compounded the problem. An industry-sponsored TCAS task force was formed under the auspices of ARINC to solve the problem. One of the solutions was a software patch in the TCAS II boxes that were already flying, so that mirror responses were less likely to occur. The software patch was called Version 6.04.

"Are you still seeing instances of mirrored TCAS maneuvers and other similar glitches? I thought Version 6.04 corrected that," I asked one LAX TRACON controller.

"6.04 did a lot to correct many of the TCAS problems," the controller answered. "They aren't *all* gone, but people (on both sides of the microphone) are getting used to the equipment. Like I said before, I think TCAS is a good idea if used correctly. I believe that in the approach environment, TCAS should be in the TA [traffic alert] mode only. It's an outstanding tool for increasing situational awareness in the cockpit, but I've seen an awful lot of crews use it as a substitute for looking out the windows. Additionally, I've seen several pilots who don't bother to decrease their vertical speed to 500 [feet per minute] in the final 1000 feet of a climb or descent which can cause the [TCAS] unit to get "nervous" in an enroute environment.

"It's a good piece of equipment, but the comments [I've been hearing] seem to indicate [that the] dozens of mid-airs [TCAS has] averted point to an ATC system that has already fallen apart. I have a hard time subscribing to that theory! In any case, the basic statement is that a controller won't sit back and count on TCAS during a [ground-based] equipment failure. He or she will use any possible means to communicate with one or both aircraft in that kind of situation."

Another forum participant asked the LAX controller:

"I flew into the LA area last month, and from the traffic that I saw in the air around me, I must ask 'what constitutes a near miss in that air space?' What are the separation rules in the [Los Angeles Basin airspace]? From [my] window seat in steerage, the other aircraft seemed close enough to touch."

To which the controller responded:

"They are the same as everywhere else." In response to the query about what constitutes a near miss in LA Basin airspace, he replied, "Paint exchange! No, really . . . most of the time, aircraft have each other in sight and visual separation is being applied. In fact, one of the more amusing pilot/controller exchanges between SoCal TRACON and an [aircraft arriving in the LAX area] was:

ATC: 'Follow a Southwest 737 at two o'clock, three miles.'

Pilot: 'Approach, I see *six* Southwest 737's at two o'clock!' "

To underscore his confidence in the level of safety he perceives in the LAX area, the controller went on to point out that there are four parallel runways in use at LAX. "They use them all at the same time! Enough said!"

Free Flight

A fundamental rethinking about the national airspace system and the ATC system has been underway for the past several years. The concept, which is called Free Flight, is being driven by rapidly changing technological, economic, and competitive forces.

The concept of Free Flight is simple. Under current "positive control" air traffic management system rules, controllers direct pilots as to when they take off, what altitudes and routes they will fly, and what speed adjustments must be made to ensure the flow of traffic. Under Free Flight, however, pilots would file flight plans for the most efficient and expeditious routing, altitudes, and speeds between two points. Instead of directing flights, controllers would monitor traffic and only interfere when a safety conflict arises or equipment fails. Pilots would have the flexibility to alter their flight plans as necessary to avoid bad weather or take advantage of more favorable routing or altitudes. However, it is likely that if Free Flight is adopted as it is now conceived, operations in high-density areas will probably still resemble today's controlled operations.

Free Flight proponents believe it offers the potential to save billions of dollars each year in fuel and time savings and an opportunity to reinvent air traffic management concepts and technologies. It is for

Free Flight separation model

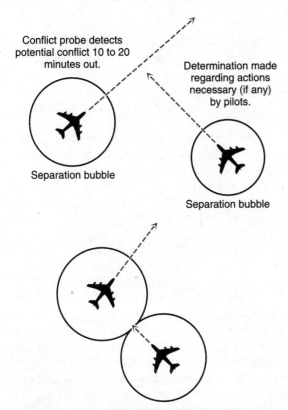

Conflict probe detects potential conflict 10 to 20 minutes out.

Determination made regarding actions necessary (if any) by pilots.

Separation bubble

Separation bubble

Figure 8.4 Under Free Flight, pilots would assume greater freedom in route planning, and currently available technology could provide the "conflict probe" traffic-separation capability that the new system would require. (RMB Associates)

this reason that Free Flight is also likely to be a key concept in many aspects of aviation computing in the next few years as the industry determines the best way to implement it.

Although there are differences of opinion regarding the best way to achieve a Free Flight environment, most of its proponents agree that Free Flight will enable pilots and operators to select the safest and most efficient routes, speeds, and altitudes in real time. It will thus combine the freedom and flexibility of visual flight rules (VFR) with full instrument flight rules (IFR) air traffic separation services.

For civil aviation authorities, Free Flight can provide a basis for renovating many outdated ATC system technologies. When equipped with current-technology navigation, collision avoidance, communications, and ATC surveillance systems, safe air traffic separation minimums can be achieved. Pilots will know where they are and where potential traffic is; air traffic managers will also know where air traffic bottlenecks are likely to occur and can take steps to recommend or implement appropriate traffic management tactics to ensure the smooth flow of traffic.

Free Flight is being given its greatest impetus by the airline industry's need to reduce the cost of flying from Point A to Point B, but also the realization by many airspace users that with a global navigation satellite system (GNSS) in place and recent advances in communications and computer technology, we now have the capability to significantly improve flight safety and efficiency and improve system capacity, while giving users more control over operational decisions and costs. Others believe that Free Flight can be achieved with the addition of low-cost, off-the-shelf software and a change in how we perceive traffic management.

To make it possible to achieve the Free Flight goals of safety and efficiency, a number of procedural and technical hurdles need to be cleared. The first hurdle is to challenge the basic notions of *controlling* versus *managing* air traffic on the part of system users, administrators, and especially air traffic controllers, whose roles and responsibilities would be recast to that of air traffic managers. Of special concern to the FAA is how Free Flight operations will fundamentally change the way aircraft are separated. Free Flight air traffic managers would likely rely on the concept of tactical separation, instead of today's strategic separation concept, which relies on projected flight paths, ATC-assigned routes, altitudes, and speed assignments.

Example of airspace availability

One cubic mile of airspace

Over 280,000 737 aircraft can be neatly parked in one cubic mile of airspace without touching.

Figure 8.5 Do we truly have an airways capacity problem? Hard to say, but this thought-provoking illustration has air traffic engineers wondering what the limits to growth might be. (RMB Associates)

Getting international consensus on such a fundamental change in thinking could prove difficult, however. For a virtually seamless air transportation system based on a Free Flight standard to be successful, both large and small nations will have to surrender at least some control of their sovereign airspace.

Two concepts

The proponents of the most widely accepted concept of Free Flight (proposed by an industry group) believe that traffic separation should be conceived of as two vertical, cylindrical-shaped zones—an outer alert area and an inner protected zone—surrounding an aircraft. The zones are defined in time parameters and displayed on an air traffic controller's "conflict probe" computer display. Information about the aircraft could be derived from computer-enhanced radar returns and aircraft-generated transponder signals containing unique, real-time data blocks containing the aircraft's identification, speed, heading, altitude, altitude trend, and other bits of information. In this context, the conflict probe would be done manually—with experienced air traffic managers who have access to various traffic management software tools. Some of these tools have been discussed elsewhere in this chapter, while others are still under development in FAA software laboratories.

Aircraft within an air traffic manager's assigned airspace would be free to maneuver at will as long as the computer and the controller know that adequate separation is ensured. If the alert zones of two aircraft touch, however, the controller can intervene and issue advisories or suggest avoidance maneuvers to the aircraft in question. The primary rule of this "game" is to prevent the inner protected zones from ever coming in contact. Assisting the controller would be the Traffic Alert and Collision Avoidance Systems (TCAS) installed on most of today's transport-category aircraft, and similar, although simpler, systems installed on virtually all small aircraft.

Another solution, proposed by Denver, Colorado-based RMB Associates, calls for a far less complex approach, using a computer-based "conflict probe." The authors of the concept state:

> The domestic airspace already has excellent communications, navigation and surveillance capability. The only new technology required to accomplish the task of enroute Free Flight is a "conflict probe." A [computerized] conflict probe uses the aircraft position and intent of the aircraft to project its path forward and determine if its protected area would overlap with that of another aircraft in a defined period of time. Obviously, the accuracy of the position and intent will provide a more accurate prediction, but radar data and flight plan intent should support Free Flight at current separation levels.

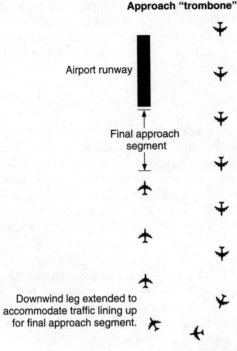

Approach "trombone"

Airport runway

Final approach segment

Downwind leg extended to
accommodate traffic lining up
for final approach segment.

Trombone can extend
50 to 60 miles before turn.

Figure 8.6 Current traffic management techniques often call for putting arriving traffic in-trail, stringing traffic out for miles, something that can be seen near major airports on any clear night. (RMB Associates)

Actually, a conflict probe is not a completely new requirement, since today's system is built around protecting the conflict probe, albeit a manual one—the air traffic controller. Today, the controller's job is to monitor an 18-inch, two-dimensional screen, mentally visualize the position of the aircraft in three dimensions and its intent five to 10 minutes into the future, and then mentally determine if a conflict (overlapping of the protected areas) will occur. As stated [elsewhere in the RMB Associates proposal], numerous layers of complexity in today's system are there simply to prevent this manual conflict probe, the controller, from being overloaded.

What is new is the requirement for an accurate computerized conflict probe which can't be overloaded. Although viewed as a monumental task, private vendors already have the capability to provide this software, off the shelf, hosted in engineering workstations. As an outcome of the Congressional hearings on our Free Flight study, [the] FAA has two conflict probe evaluations scheduled for Boston Center and Kansas City Center in late 1995 through 1996.

Admittedly, RMB Associates has focused on the what most call a "simplistic" approach to the implementation of Free Flight. This approach has grown out of attacking Free Flight implementation by asking a different question than most in relation to the problem. Many ask the question

Approach under Free Flight

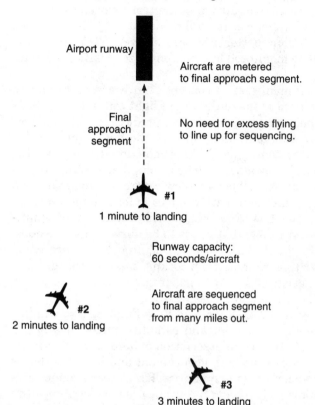

Airport runway

Aircraft are metered
to final approach segment.

Final
approach
segment

No need for excess flying
to line up for sequencing.

#1
1 minute to landing

Runway capacity:
60 seconds/aircraft

#2
2 minutes to landing

Aircraft are sequenced
to final approach segment
from many miles out.

#3
3 minutes to landing

Figure 8.7 Free Flight will enable pilots and ATC managers
to coordinate arriving traffic flows more efficiently, saving
fuel and time while reducing noise. Computers will play a
major role in Free Flight. (RMB Associates)

"How can GPS, Data Link, ADS-B [Automatic Dependent Surveillance],
TCAS, etc., be utilized to usher in a Free Flight environment." RMB Associates approached the problem from a different perspective and with a different question, "What is stopping the pilot from taking off and flying
their preferred path to the destination and, if so desired, changing that
path enroute." Both of these questions, answered correctly, can lead to the
implementation of Free Flight, but only the second question minimizes
the task. The aviation industry can not continue to throw technology at
the problem of Free Flight implementation. We must define the underlying task, safe separation of aircraft in a random path Free Flight system,
and apply the minimum technology required.[10]

Regardless of which concept eventually proves the most feasible, proponents of Free Flight management systems foresee the development

of a hierarchy of ATC procedures, ranging from free maneuvering to more traditional (restrictive) air traffic control clearances in high-density airspace. The level of control will be dictated by the "dynamic density" of traffic and flow patterns within a given ATC sector. The dynamic density will be monitored by a computerized traffic flow management system.

Obviously, some new management concepts, hardware, and software tools need to be developed to make the Free Flight concept work. Software is required to help air traffic managers predict conflicts and bottlenecks and divert traffic and, later, analyze the data. One possible piece of the puzzle is a system called PRAT (Prediction/Resolution Advisory Tool) a computer/display adjunct to host computer systems. PRAT, if it lives up to its promises, will better allow controllers to project and analyze random routes and flight path changes in four dimensions.

It remains to be seen how Free Flight will ultimately be implemented. One of the first things that needs to be answered is the question of what Free Flight is. Many aspects of Free Flight are being considered at this writing, ranging from cockpit-based versus ground-based technologies, geographic coverage, human factors, OEM/airline/user group/FAA roles, and other issues.

The formation of an RTCA task force on Free Flight is likely to further refine the Free Flight concept and establish milestones for its implementation. RTCA, Inc., is an association of aeronautical associations of the United States from both government and industry that is dedicated to the advancement of aeronautics. RTCA seeks sound technical solutions to problems involving the application of electronics and telecommunications to aeronautical operations. Its objective is the resolution of such issues through consensus. The findings of RTCA are usually recommendations made to all organizations concerned. Since RTCA is not an official agency of the United States government, its recommendations shouldn't be regarded as statements of official government policy. However, when the FAA develops new certification requirements, it leans heavily on RTCA recommendations.

As this chapter was being written, FAA Administrator David Hinson formalized the agency's support with the creation of a Free Flight steering committee "almost immediately," but he declined to say at what level the FAA will participate. The committee is vital to beginning the Free Flight process, and the RTCA task force recommended in October 1995 that it be cochaired by the FAA administrator and an airline executive officer. Robert Baker, executive vice president of operations for American Airlines, and other members of the RTCA task force have stressed the importance of participation by all industry elements. Baker said the Free Flight concept, "will die without a full collaboration of all stakeholders, including FAA." John O'Brien, safety director for the Air Line Pilots Association (ALPA), said the only concern of pilots is that the effort "be conducted at the highest level to gain all the benefits."

The role of the steering committee, as defined by the RTCA task force, would be to establish an implementation strategy and milestones. It also will review progress periodically and identify new implementation opportunities and events or situations that inhibit progress. Committee members will represent aviation groups, unions, and the Defense Department.

RMB Associates, which has proposed a simpler means of implementing Free Flight, argues that the current RTCA task force lacks sufficient leadership and will likely fail to make swift, meaningful implementation progress. Indeed, some observers within the FAA don't foresee Free Flight being in place before 2010.

Nevertheless, the Free Flight effort has been endorsed by the Air Transport Association (ATA), the National Business Aircraft Association (NBAA), the Aircraft Owners and Pilots Association (AOPA), the Air Line Pilots Association (ALPA), and the National Air Traffic Controllers Association (NATCA).

"The challenge is implementation," RTCA president David Watrous said. "Many problems still are to be solved, including . . . the military's special-use airspace." O'Brien said much of the technology needed for Free Flight will be on the ground, and the "ultimate responsibility for separation [of aircraft] will remain on the ground." Baker warned that without Free Flight, the expense of dealing with future congestion will drive up costs for airlines and passengers.[11]

National Route Program and other FAA initiatives

The FAA hasn't been completely without its own ideas to improve airspace safety and efficiency. For example, with the National Route Program (NRP), initiated in the United States in 1992, the FAA began removing restrictions to flight at higher altitudes, commencing with Flight Level 330 (FL330), or 33,000 feet and up, for airspace west of the Mississippi River and FL350 for airspace east of the Mississippi). The aim is to allow more route flexibility—and thus greater efficiency—for the airlines. The next block of airspace to be added to the NRP will bring it down to FL290.

The FAA estimates industry-wide NRP-related annual savings in fuel and time to amount to some $40 million. Some critics claim the program is too restrictive to make significant gains in time or fuel savings. A follow-on program, Expanded NRP (ENRP), removes the FAA's city-pair and leg-length restrictions from NRP. In a nutshell, ENRP applies to flights operating at or above FL330 and allows user-preferred routes from a point 200 miles from the departure airport to 200 miles out from destination. As more unrestricted airspace is added to the program, more users will benefit and savings will grow.

Another program, called RVSM, for Reduced Vertical Separation Minimums, will commence on January 2, 1997. This program will, as the name implies, reduce the vertical spacing requirements for jet traffic in certain areas of the North Atlantic to increase capacity. Currently, aircraft operating in what is called the MNPS (minimum navigation performance standards) airspace at altitudes between FL290 and FL410 are separated vertically by 2000 feet. In 1997 that will be reduced to 1000 feet, provided aircraft operating there meet new RVSM requirements, which basically call for two independent altitude-measuring systems, an altitude-alerting system, a second surveillance radar altitude-reporting transponder, and an autopilot that can maintain altitude within 65 feet.

In trials on transoceanic routes with prior approval from air agencies, selected air carrier aircraft are being allowed to maintain lateral and vertical separation from other flights by means of onboard TCAS equipment.

FAA Modernization Efforts

The FAA has initiated a number of modernization efforts since its inception. Critics allege, however, that most of the recent efforts have been unsuccessful and squandered millions of dollars in taxpayer money. Despite some sparkling exceptions, the agency's modernization plans seem to be hamstrung by conflicting priorities, a string of short-term administrators, lack of oversight, and other problems. Nevertheless, the FAA completed 10 of 16 projects planned to be completed in 1994; reasons cited for noncompletions were software problems and site-preparation delays.

The FAA's estimate of the total cost to modernize has decreased. In its December 1994 financial plan, the agency estimated the cost of upgrading its ATC system from 1982 through 2003 will cost $37.3 billion, down from $38.8 billion the year before. Of the current estimate, $19.8 billion was appropriated from 1982 to 1995. The financial plan projects that $17.5 billion will be needed from 1996 through 2003. However, because of anticipated funding cuts, the FAA is currently reviewing the funding levels for its laundry list of 158 projects, and the agency might request funding for $14.7 billion over the same period, which will likely require the elimination of some projects and funding extensions to others.

Air route traffic control centers

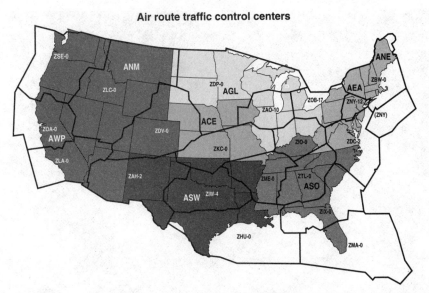

Figure 8.8   The FAA currently operates 21 Air Route Traffic Control Centers (ARTCCs). These feed data into the Central Flow Control facility in Herndon, Virginia. (FAA)

AAS

The centerpiece of the agency's most recent modernization effort was a program called the Advanced Automation System (AAS). It was originally conceived in the wake of the 1981 air traffic controllers' strike and subsequent firing of about 11,000 controllers. Air traffic was booming, following the Airline Deregulation Act—something needed to be done to prevent the ATC system from being gridlocked.

As originally designed, AAS would replace existing display and computer systems. It also would have enabled the FAA to consolidate 203 TRACONS and 20 ARTCCs into 23 Area Control Facilities (ACFs) with five basic tools:

- The Peripheral Adapter Module Replacement Item (PAMRI), a communications computer that connects en route Centers with external systems, such as radar

- The Initial Sector Suite System (ISSS), which would have replaced existing controller workstations (ISSS was later scaled back and became Display System Replacement, or DSR)

- The Terminal Advanced Automation System (TAAS), which would have provided TRACON controllers with workstations similar to ISSS

- The Tower Control Computer Complex (TCCC), designed to replaced hardware and software at selected control towers

Figure 8.9 An air traffic controller working with the Voice Switching Control System at one of the FAA's Air Route Traffic Control Centers (ARTCC). The agency's computer upgrades have been beset with growing pains and costly overruns, but new initiatives appear to be mostly on-track. (FAA)

- The Area Control Computer Complex (ACCC), a series of hardware and software enhancements and integration efforts to replace the existing host computer and build upon the ISSS equipment in the ARTCCs to allow for their conversion to ACFs. ACCC was later eliminated.

AAS also included advanced Automated En Route ATC (AERA) software that is expected to give controllers the tools to identify and resolve potential airspace conflicts. Only PAMRI—the least complex of all the tools—was completed.

In 1988, the agency selected a proposal from IBM to develop AAS. (Loral later purchased the IBM division that was building AAS.) Shortly after, design and development problems appeared, and delays in the program began.

The total cost estimates of this program grew from $2.5 billion in 1983 to $7.6 billion in 1994, and the project slipped eight years from its original schedule. In 1992, Congress told the FAA to start looking into alternatives. A limited consolidation plan was developed later that year. Several of the program's major components were revised or scaled back. However, by late 1993, the FAA determined that the program's costs would increase to $5.3 billion. Internal agency studies indicated the likelihood of further schedule slippage and mounting costs. A second study, conducted by the Center for Naval Analyses, recommended that the FAA revalidate the AAS program requirements. Another FAA study developed alternative strategies for the program.

In June 1994, on the basis of these studies, FAA Administrator David Hinson decided to eliminate some $500 million in software from the program and authorized a more practical and less-costly version that would deliver approximately 1000 new ARTCC controller workstations. Additionally, a new FAA organization for air traffic systems development was created.

The AAS program's restructuring has resulted in the creation of three more manageable projects that are estimated to cost some $6 billion, all told. DSR is expected to begin replacing en route hardware and software by September 1998. Loral will supply the DSR system. The agency also hopes that DSR will become a platform for more advanced capabilities.

After the FAA implements DSR, it plans to introduce an initial AERA capability in 1999. Additionally, the agency added a Display Channel Complex Rehost project, which is considered to be an insurance program that will execute older software on a new commercial off-the-shelf computer in the event that DSR cannot be fielded before the existing display channel reaches a critical point.

In 2001, the FAA also will add an enhanced Direct Access Radar Channel (DARC) replacement project to provide an up-to-date automation backup infrastructure to all ARTCCs. In 2002, the host computer is scheduled to be replaced to increase capacity.

According to the General Accounting Office (GAO), the ATC system that emerges from the original AAS concept will have less capability than was originally required. Development of the various elements of the system continue to be plagued with software development problems.[12] Additional ATC concepts related to data link and Future Air Navigation System (FANS) implementation efforts are detailed in Chapter 10.

Other FAA R&D

Some other examples of FAA research and development programs include:

- *TFM.* Traffic management specialists at the ATCSCC and at traffic management units at major ATC facilities are responsible for traffic flow management (TFM)—the process of directing dynamic traffic flows using limited NAS resources. The FAA's research and development (R&D) community developed the traffic management system—a set of automated decision support tools used for balancing air traffic demand and NAS capacity.

The core of the automated system is known as the Enhanced Traffic Management System (ETMS), which is discussed elsewhere in this chapter. Traffic management specialists also use other stand-alone air traffic management systems, such as the High Altitude Route System (HARS) and Ground Delay Manager (GDM).

The FAA R&D community developed these systems using a broad model for problem solving—by gathering information and gaining situational awareness, identifying problems, developing solutions, implementing the approved design, and performing operational performance reviews. At each step, analysts and computer programmers worked with operations personnel who will ultimately depend on the system.

Current research focuses on evolving the current TMS into a more flexible, collaborative system that better considers the airlines' preferences. Current development focuses on migrating the system to an open system architecture and decoupling applications from data and infrastructure to allow rapid prototyping. The FAA R&D team consists of a handful of small, specialized teams working in concert and orchestrated by a unifying systems engineering effort.[13]

- *FADE/SIM.* The FAA-Airlines Data Exchange/Simulation (FADE/SIM) offered traffic management specialists an opportunity to understand how frequent schedule updates provided by airlines would affect TFM decision-making. Using computer-based simulations, TMSs can effect Ground Delay Programs as needed in realistic situations if up-to-the-minute airline schedule changes are available. Results are recorded for analysis of how their decisions might be different in light of up-to-the minute schedule information.

  Using the FADE simulation, TMSs assessed the effects of projected demand, changing traffic acceptance rates, making program revisions, and running other scenarios as updated schedule information is received on the fly from the carriers.[14]

- *HARS.* The High Altitude Route System is an automated traffic management tool that determines optimal flight routes based on aircraft performance, changing weather conditions, traffic demand, and resource limitations. HARS predicts what airlines will file for flight plans and produces alternate route strategies around severe weather areas, special-use airspace, or congested sectors.

  HARS was used as an operational prototype by ATCSCC specialists to find optimal routes by creating aircraft-specific profiles and combining traffic schedules with airspace data, dynamic weather forecasts, and severe weather information. Specifically, HARS mathematically determines distances, navigation parameters, and en route travel times to provide the following capabilities:

- Generation of route revisions as changes in weather, user demand, or system capacity occur
- Anticipation of route requests for major city-pairs in support of the FAA's National Route Program (NRP), which is discussed elsewhere in this chapter, and nonpreferred route requests

Using this data, say FAA officials, traffic flow managers can—in real time—optimize the use of NAS and reduce time and fuel costs for aircraft operators. By using HARS-generated profiles to anticipate airspace user requests, the ATCSCC can provide efficient traffic flow management strategies. HARS significantly improves the management of direct routings, severe weather avoidance routings, and nonpreferred route requests.[15]

- *GDM.* Ground Delay Manager is a software display tool that allows TMSs to evaluate the effects of ground stop or ground delay programs (when ground controllers don't allow aircraft to leave the gate at affected airports until up- or downstream air traffic bottlenecks are cleared). It can be used either to simulate a revision to a program that is in effect or simulate various release scenarios for a current ground stop. It can also be used to monitor the progress of programs.

  Using GDM, traffic specialists can monitor whether the number of predicted arrivals at a controlled airport is running above or below the rate that they are trying to achieve. Looking at a chart of predicted arrivals (displayed intervals updated every five minutes), specialists can compare the predictions against an airport acceptance rate and other data. If a ground stop isn't producing the desired results (or if the rate changes) specialists can use the Delay Manager to release flights, thereby experimenting with different ways to resolve the problem.

  TMSs use Ground Delay Manager to determine the best flights to stop, which flights to release, and when to release flights. The GDM shows what is likely to happen. GDM displays its results graphically and operates in its own window on the ETMS system, so specialists can use it independently of other traffic management tools. Any request for information can be made through a point-and-click interface, with results displayed in seconds.[16]

- *Airport Lighting Control Monitor.* ALCM is a touch-screen airport lighting display and control interface for use by FAA personnel at the new Los Angeles International Airport control tower cab, which was dedicated in March 1996. The system enables controllers and technicians to monitor and control LAX's many runways, ramps, and taxiways using touch-sensitive screens that graphically depict the location of lights throughout the airport. LAX's lighting configuration and requirements are particularly complex because the airport has four parallel runways located in two separate complexes and interconnected by more than two dozen taxiways. The system, provided by Systems Resources Corporation, a Burlington, Massachusetts, systems engineering and integration company, features industry-standard computers, fiber-optic communication, UNIX servers, and programmable logic controllers. The system is incrementally expandable so it can keep pace with LAX's future runway/taxiway lighting requirements. The system it replaced relied on manual toggle switches and voice communications with ground vehicles and aircraft to verify the airport's lighting status.

These examples of FAA programs (as well as those under development at the FAA Tech Center described later in this chapter) also serve another purpose here. They provide a glimpse at the types of operational challenges and solutions that face ATC specialists, system programmers, engineers, and managers. With the eventual development of entirely new paradigms of air traffic management, such as the National Route Program and Free Flight, the need for imaginative, informed software programmers and analysts and computer hardware developers is almost certain to increase. As pointed out in Chapter 2, there is a significant shortage of aviation-minded individuals with computer-related career aspirations. ATC management and system development might be one area to consider. The agency is also committing itself to using commercially available off-the-shelf (COTS) computer systems whenever possible, rather than spending taxpayer dollars to develop new systems, thus, private-sector software system developers may also benefit.

FAA Tech Center: Where the wheels leave the runway

One of the most fascinating places in "ATC World" is the FAA's Tech Center, located at the agency's sprawling campus adjacent to Atlantic City Airport, near (no surprise) Atlantic City, New Jersey. Although it has been several years since I've had an opportunity to visit the place, I've tried to keep up with what's going on there.

The Tech Center comprises a number of systems R&D and support divisions and subdivisions, including

- ATC
  - ATC Engineering and Test Division
  - En Route and Terminal ATC Labs

- National En Route Systems Engineering Division
- Aviation Simulation & Human Factors
  - Human Factors Laboratory
  - FAA Human Factors
  - National Simulation Capability
  - Target Generator Facility
  - Virtual Environments & Advanced Visualization Laboratory
  - Reconfigurable Cockpit Simulator Lab
- Airport & Aircraft Safety
  - Airport & Aircraft Safety R&D Division
  - Aviation Safety Program
  - Precision Runway Monitor (PRM) System Final Monitor Aid (FMA)
  - Cabin/Fire Safety Labs
- Aviation Security R&D

To be fair, if the FAA is thinking about developing a new information management system, it is likely to conceived at MIT's Lincoln Laboratory in Cambridge Massachusetts, where it can draw upon the brain power at the Massachusetts Institute of Technology or at nearby Mitre Corporation. But it is the Tech Center where the "wheels leave the runway," so to speak, as dazzling new ATC concepts make the leap from theory to reality. In the specifications and development of the highly automated ATC systems of the near future, the Tech Center plays a major role and furnishes support for current systems as well.

The four-story, $150-million Computer Complex at the Tech Center houses the development testing area. There, the latest hardware and software are tested in darkened rooms simulating the FAA's ATC facilities. Adjacent rooms contain banks of mainframe computers, while other rooms contain pleasant work areas for computer programmers, airspace engineers, technical researchers, computer technicians, systems analysts, and research psychologists.

During my last visit, the programmers and engineers in the Aircraft & Safety Systems Group I met were struggling with several interesting projects. One of them was the Initial Sector Suite System (ISSS), the state-of-the-art controller workstation that was supposed to replace existing hardware and software that had been developed in the 1940s and 1950s. (ISSS was later scaled back and became the DSR program.)

Another program, which ultimately did grow legs and find a home, is the Precision Runway Monitor (PRM) program. Although the program development took place several years ago, it is a typical example of the types of programs the Tech Center tackles. Later, I briefly describe some of the Tech Center's current programs.

Precision Runway Monitor. The Precision Runway Monitor (PRM) program was developed to handle the expected high-density and high-frequency arrival traffic at airports with closely spaced, parallel runways. Airports such as Dallas/Fort Worth, Atlanta's Hartsfield International, JFK International in New York City, Los Angeles International, and the new Denver International were designed and built to use parallel runways to handle their traffic—sometimes with as many as four parallel runways.

The FAA controller's rule book dictates that when runways are spaced—centerline to centerline—less than 4300 feet apart, approaching aircraft must be staggered in their sequencing to avoid creating horizontal conflicts. This requirement reduces the efficiency of some parallel runways, so the Tech Center was tasked with implementing procedures, hardware, and software to make better use of these runways while maintaining adequate levels of safety.

The PRM program was launched in the 1980s when the FAA began to look at ways to digitize radar displays and increase the speed of radar sweeps. At about the same time, the FAA's Southwest Region wanted to develop procedures to enable three or even four simultaneous "formation" ILS approaches at airports having multiple parallel runways, such as at Dallas/Fort Worth. AlliedSignal and General Electric had developed radars to make the program feasible and had installed demonstration programs at Memphis Regional in Tennessee (MEM) and Raleigh-Durham Airport (RDU) in North Carolina.

To make full use of PRM, an new TRACON controller position was created—the final monitor controller (FMC). The program calls for one FMC per runway. The FMC's sole responsibility is to ensure that none of the other aircraft on the adjacent approach paths (or overflights) strays into the approach path airspace. If another aircraft "blunders" into the FMC's designated "no transgression zone," the FMC can override the local controller's frequency and issue instructions to the affected aircraft.

The term "blunder" is used loosely and isn't meant to disparage pilots' abilities. It only refers to the numerous traffic variables involved in managing high-density traffic. A combination of events that wouldn't happen on a single approach path could occur through nobody's fault and quickly snowball into a risky situation if it weren't for the PRM concept and the related display technology.

The mayhem simulator. The PRM concept was being tested during my visit to the Tech Center. The Computer Complex laboratory used extensive simulations involving a bank of PRM displays manned by "FMCs" in one room; another room was simulating a "local control" position, and another room contained "pilots." In yet another room was a bank of computers that were simulating the movements of up to 425 "aircraft." These aircraft were being "controlled" from yet another room, where up to 36 specially trained personnel responded, via computer keyboards and headsets, to realistic Center and TRACON instructions.

Meanwhile, by prior arrangement, seven major air carriers were conducting approaches in high-fidelity flight transport simulators at facilities located around the United States. For added effect, the simulators were connected—in real time—to the PRM computers in the Tech Center Computer Complex via high-speed modems. One simulator, mimicking the flight characteristics of a slow, piston-engine light twin—an airplane that would typically fly slower than the Boeings, Airbusses, MD-80s, and Lockheeds—would occasionally be a fly in the ointment.

The mix of traffic appeared to flow smoothly inbound to three of the parallel north/south runways at a simulation of the new Denver International Airport in Colorado, which was still under construction at the time. Suddenly, a data tag attached to a Boeing 727 on a 2-mile final to Runway 17C (the middle runway) began to drift off the approach path centerline into the path of a Japan Air Lines Boeing 747 approaching Runway 17R.

"United 22, go around," said the FMC working Runway 17R traffic, stepping on the local controller's frequency. The crew of the Boeing acknowledged the controller's request, and its data tag responded accordingly. *Der Blundermeister* had struck again. It was a traffic specialist located in yet another room in the lab, whose task it was to interject preplanned mayhem to find the limits and weaknesses of the Denver PRM and the FMCs. The blunder specialist directed the scenarios while associates took notes and recorded the events on a high-tech videocassette recorder for later playback and analysis.

National Simulation Capability.   The Tech Center's National Simulation Capability is a division working in another promising area of computer technology. NSC is one element of the Tech Center's Aviation Simulation and Human Factors Division's Simulation and System Integration Branch.

In 1988, Congress recognized the role of simulation in the research and development process and expressed that recognition through legislation. The Aviation Safety Research Act of 1988 required the FAA to ". . . undertake . . . a research program to develop dynamic simulation models of the air traffic control (ATC) system . . . which will provide analytical technology for predicting airport and ATC safety and capacity problems, (and) for evaluating planned research projects." The NSC program was established in response to that congressional mandate.

The NSC program, originally titled National Simulation Laboratory (NSL), began in 1990 under the concept of creating a stand-alone simulation facility to satisfy this mandate. In 1991 a cost-effective decision was made to utilize existing resources in a distributed simulation network environment, hence the name change to NSC.

The NSC is a large-scale, distributed, simulation network comprised of various air traffic-related simulators and highly skilled personnel. The NSC is used to facilitate the development of the future National Airspace System (NAS) by providing an integrated representation of the NAS through the interconnection of various facilities and laboratories internal and external to the federal government. The mission of the NSC is twofold. First, the NSC creates a synthetic, virtual representation of the future NAS, connecting separate subsystems and components. Second, the NSC systematically examines future operational concepts and technologies for the NAS as a whole. In doing so, issues of component integration and cross-component interaction are discovered that otherwise would not surface until much later in the development process.

Virtual ATC. The Tech Center's Virtual Environments & Advanced Visualization Laboratory has developed a state-of-the-art capability for conducting work in the areas of virtual environments and advanced visualization, also known as virtual reality (VR). Virtual reality has been popularized in the mass media through movies such as *Lawnmower Man, Speed,* and other sci-fi adventures. In the Tech Center, VR takes on a new, if more practical meaning.

Virtual reality is a concept for computer-human interfaces that uses new technologies and techniques and is more natural and effective than the typical existing interfaces. VR allows users to dynamically interact with a 3-D graphical representation of concepts, designs, and data sets that might otherwise be too complex to visualize. Users can be visually and aurally immersed in a virtual environment that might represent some set of complex data. The level of immersion can vary depending on the equipment used, such as head-mounted-displays, earphones, multiple large-projection screens, and so on.

Operational concepts being explored in the VR lab include

- Proposed new concepts such as aircraft approach technologies and air traffic flow patterns
- Architectural design proposals such as airport, control tower, and other ATC facility layouts
- The ergonomic design of air traffic control consoles; data sets of flight recorder data for accident reconstructions; weather in an airspace;

luggage going through a scanner; stress being applied to various parts of an aging aircraft; and maintenance diagrams and information.

Also being developed is a low-cost simulator used for a virtual control tower. This alternative to large-scale simulators costs less to develop, maintain, and operate and is more versatile, which expands the number of its uses. The folks in the VR lab are also studying areas within the FAA in which the application of VR would improve the effectiveness and efficiency over methods currently being used to study data, designs, or concepts, as well as proof-of-concept studies for possible air traffic systems using VR.

Fertile Ground

Criticism of the FAA has become something of a cottage industry, particularly inside the Washington, D.C., beltway. It is easy to fault the system for being obsolete and the fixes for being more costly than anticipated. But compared to what? Europe? Russia? China? As one of the supervisors who escorted me around the Tech Center said, "It's not as if you can go to a store and buy what we're building here."

The folks at the Tech Center aren't resting on their laurels, but neither do they take the criticism to heart. They're too busy making sure the U.S. ATC system continues to be the best in the world.[17]

As I said at the outset of this chapter, the material I've presented here is neither an indictment nor an endorsement of the FAA, its programs, or any of its employees. The information management challenges the FAA is facing today are formidable, but I am very optimistic that it will find solutions to those challenges in the near future.

What form the FAA's ATC system will ultimately take to meet the safety, budgetary, and management requirements of the NAS and its users is obviously still open to debate, consensus and an awful lot of development. The task is already drawing upon the energies and talents of many outside the FAA and will continue to do so for some time. This is why I think that ATC remains one of the most compelling subjects in aviation computing—and also one of the most fertile grounds for individuals with imaginations to pursue. ATC holds, in my humble estimation, the area with the greatest number of career possibilities, whether in the government or in the private sector.

References

1. Air Transport Association. 1996. *ATA Handbook.*
2. *Aviation Daily* item. 1995. October 1.
3. Business & Commercial Aviation. 1996. "ATC On the Blink" January.
4. FAA. 1996. Document describing the ATCSCC, supplied by FAA Public Affairs Office. March.

5. FAA. 1996. National Airspace System Architecture, Version 1.5. February.
6. Air Transport Association. 1996. *ATA Handbook.*
7. *Aviation Week & Space Technology.* 1996. "NTSB Report Declares ATC System Safe" January 22.
8. RMB Associates. 1996. *The Business of Free Flight.*
9. FAA. 1996. National Airspace System Architecture, Version 1.5. February.
10. Baiada, R. Michael. 1995. *Free Flight—Rapid Implementation.* RMB Associates.
11. *Aviation Daily* item. 1996. March 18.
12. GAO Report on Air Traffic Control. 1995. "Status of FAA's Modernization Program" May.
13. FAA Systems Engineering & Development. 1994. *Developing the Traffic Management System: Concurrent Engineering Within the FAA R&D Community.* Federal Aviation Administration. August.
14. FAA Systems Engineering & Development. 1994. *FAA/Airlines Data Exchange/Simulation.* Federal Aviation Administration. August.
15. FAA Systems Engineering & Development. 1994. *High Altitude Route System.* Federal Aviation Administration. August.
16. FAA Systems Engineering & Development. 1994. *Ground Delay Manager.* Federal Aviation Administration. August.
17. *Business & Commercial Aviation.* 1993. "The FAA's Tech Center." March.

The Weather Net and Uncle Sam Online

How do you keep track of 150,000 pilots and thousands of aircraft? How do you make sure civil and military flight crews have the latest weather forecasts? How should you communicate with an agency made up of tens of thousands of employees in 12 regional offices and a couple dozen other facilities?

Although ATC gets a lion's share of the FAA's computing budget, other parts of the FAA and other federal agencies with aviation-related information technology systems are also important elements of the aviation computing picture. One of the largest segments is the government's weather service providers. We take a look at how weather information is gathered and exchanged within the United States and beyond. Additionally, we examine some of the more promising areas of information systems development in weather forecasting and dissemination.

The Big Weather Picture

By the year 2000, government sources predict that some 800 million revenue passengers will use the U.S. air transportation system. Additionally, tens of thousands of others will use private, corporate, and government aircraft for a variety of purposes. Since weather accounts for a substantial percentage of air traffic delays, improvements in aviation weather services are essential. Fortunately, new concepts and technologies are beginning to take advantage of new weather gathering, forecasting, and disseminating capabilities that might provide pilots and others with more timely and useful weather information. Promising new weather sensors are coming on line, and others are in development. Improvements in macro-, meso-, and microscale weather analysis and forecasting models are being made possible by advances in computer technology and weather science.

In this section, we take a look at the basic structure of the aviation weather system and some of the promising new technologies that are under development to make weather products that pilots use easier to obtain, more timely, and accurate. Four government agencies have major roles in providing and distributing aviation weather information. These agencies are within the Departments of Commerce (DOC), Transportation (DOT), and Defense (DOD).

There has long been a need for an integrated, interagency strategy to ensure that the U.S. aviation weather system evolves to meet the operational needs of national airspace users and the ATC system as it evolves. To meet this need, the National Aviation Weather Program Plan (NAWPP) was developed in the 1980s under then-FAA Administrator J. Lynn Helms. NAWPP was implemented to create a vision of future weather products and help define the requirements of an integrated, interagency weather system. At the same time, the products that emerge from the weather science labs must be capable of meeting international standards for aviation weather products and services. Although the NAWPP effort is ongoing (and I discuss it again later in this chapter), delays have slowed its progress. Nevertheless, some elements of the plan have become part of the aviation landscape.

The National Weather Service (NWS), along with other branches of the National Oceanic and Atmospheric Administration (NOAA) under the DOC, is the primary provider of weather services to the public and other government agencies. The NWS is the domestic coordinator in a worldwide data acquisition network in which the FAA, DOD, private observers, and international agencies all participate. Observation data collected through this network include surface, upper air, radar, and satellite-based weather information. NWS specialists integrate and analyze the various types of observations to generate a wide range of products, some designed for general public use and some tailored for special uses such as aviation, agriculture, marine, environmental, and forestry applications. In disseminating these products, the NWS cooperates with other government agencies and with the private media. A variety of manual and automated sensors are used to gather the weather.

The DOD provides global meteorological services tailored to meet the unique requirements of the military services. The U.S. Air Force supports Air Force and Army operations, and the U.S. Navy Naval Oceanography Command (NAVOCEANCOM) supports Navy and U.S. Marine Corps operations. In performing their respective missions, the Air Force and NAVOCEANCOM perform many of the basic meteorological functions of the NWS as a basis for the specialized services they provide.

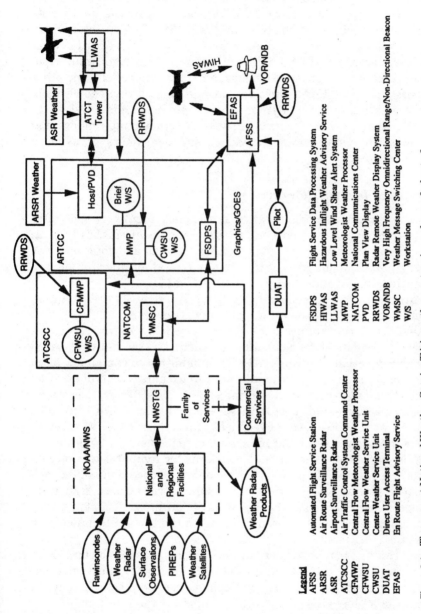

Figure 9.1 The current National Weather Service/FAA weather system is made up of a host of interrelated programs. This diagram does not show the National Aviation Data Interchange Network, which is, in turn, connected to the Aeronautical Fixed Telecommunications Network. The AFTN is an international network of weather air traffic messaging systems. A new satellite-based system called World Area Forecast System has just been inaugurated. (NWS)

Legend

AFSS	Automated Flight Service Station
ARSR	Air Route Surveillance Radar
ASR	Airport Surveillance Radar
ATCSCC	Air Traffic Control System Command Center
CFMWP	Central Flow Meteorologist Weather Processor
CFWSU	Central Flow Weather Service Unit
CWSU	Center Weather Service Unit
DUAT	Direct User Access Terminal
EFAS	En Route Flight Advisory Service

FSDPS	Flight Service Data Processing System
HIWAS	Hazardous Inflight Weather Advisory Service
LLWAS	Low Level Wind Shear Alert System
MWP	Meteorologist Weather Processor
NATCOM	National Communications Center
PVD	Plan View Display
RRWDS	Radar Remote Weather Display System
VOR/NDB	Very High Frequency Omnidirectional Range/Non-Directional Beacon
WMSC	Weather Message Switching Center
W/S	Workstation

The FAA, as part of the DOT (for the time being), is primarily a user of the services provided by the NWS and DOD. However, weather observations and pilot reports (PIREPs), collected by the FAA for its own use are distributed to the NWS and DOD. The FAA's primary function is distributing aviation weather information to the aviation community, including civil and military pilots, air carriers, municipal agencies, corporate flight departments, air traffic controllers, and other FAA personnel. The FAA is also responsible for fostering the development of aviation-specific weather sensors, products, and services. In addition, other federal agencies, as well as commercial vendors and state aviation agencies, are providers and consumers of aviation weather services.

International Weather Networks

Since weather knows no boundaries, it became evident very soon after pilots began to fly transcontinental and transoceanic flights that an international network of reliable weather reports and forecasts was necessary. Much of our current international weather information was constructed during and shortly after World War II.

I contacted Jerry Uecker, at the National Weather Services Aviation Branch, to explain this sometimes arcane network:

"The network that connects the U.S. ATC system with the rest of the world's aviation telecom networks is called the Aeronautical Fixed Telecommunications Network (AFTN). In addition to carrying a lot of ATC information, the AFTN is also used to exchange a lot of aviation weather. That system runs at varying speeds, and is at various stages of modernization—and decrepitness—around the world. In some places it's at a 1960's level of technology; in others it's in the 1990s. Much of it still uses telex, but more and more it's using X.25 protocol."

X.25 is one of the international standards for computer telecommunications. If you've ever used one of the online service companies, such as America Online or CompuServe, for example, you've used X.25. Data is transmitted at speeds that range from a snail's-pace 75 bits per second to a blistering 2400 Kbps, depending on the level of technology in the area.

"Each country's civil aviation authority is responsible for maintaining and upgrading their weather and ATC [telecommunications] network in coordination with others," said Uecker. "Of course, a lot of countries don't have the wherewithal to upgrade their systems."

In the United States, the FAA depends on the National Aviation Data Interchange Network (NADIN) to exchange weather and other operational information. NADIN has two hubs, one in Salt Lake City, Utah, and the other in Atlanta, Georgia.

"Prior to the establishment of NADIN, there was the Kansas City Weather Message Switching Center (WMSC)," explained Uecker. "That has been incorporated, to one degree or another, into the hubs."

The two hubs, in turn, are connected to the Central American Meteorological Network (CMET), The Antilles Meteorological Network (ANMET), Bermuda, Canada, Mexico, and beyond. The AFTN is fairly robust and modernized in western Europe, but is in varying stages of disrepair the further east and south one travels. Major trunks link the AFTN system to Japan, Australia, China, the Indian subcontinent, parts of the Middle East, Africa, and South America. Some links are subject to bilateral agreements, while other circuits are still only in the planning stage.

The information is exchanged on circuits within the network on a need-to-know basis. Obviously, a weather station in Tashkent or Pago Pago wouldn't have much use for weather in Bangor, Maine—or vice versa. (If, however, you were in need of that information, a private weather vendor such as Universal Weather and Aviation or Jeppesen could track it down for you.) Uecker pointed out that even in the darkest days of the Cold War, weather information continued to be exchanged between China, Russia, and the West.

Within the United States, weather and other aeronautical information is also distributed to the air carriers and other users (such as DUAT) on what's called the 604 Circuit and by other high-speed lines. The FAA exchanges weather directly with the NWS's Silver Spring, Maryland, headquarters. Private weather vendors and Flight Service Stations tap into FAA weather circuits as well as NWS circuits. "The FAA drives a lot of the weather circuits in the United States," said Uecker.

The NWS also produces a fair share of the information meteorologists use. One major system produces computer-modeled weather data from NWS supercomputers located in Suitland, Maryland, to end users at 56 Kbps. This information is consumed by a variety of end users, including the military and the Global (Meteorological) Telecommunications System (GTS). The GTS is an equivalent to the AFTN.

"While the AFTN uses only bits of weather information to produce a forecast for a location," said Uecker, "the meteorologists need lots of information from all around a given location—upstream and downstream, so to speak. Thus the data requirements are much greater for meteorological aspects than they are for aeronautical aspects."

WAFS: A change in the wind

The ICAO and the World Meteorological Organization (WMO) have been the prime movers behind a program called the World Area Forecast System (WAFS). The WAFS program, which went into service in 1995, has two major components. One is meteorological, and the other is a promising new communications system.

The communications portion is operated and largely funded by the United States and the United Kingdom. Each operates a World Area Forecast Center, and each system backs up the other. Alden Electronics won the contract to develop the computer processors for the system, and MCI and Hughes operate the communications portion of the system, according to Uecker.

The information that is produced—aviation weather products (global flight-planning data such as winds, temperatures, TAFS, and METARs)—is disseminated at 38.4 Kbps by a trio of geosynchronous Intelsat satellites. The system has been happily received, especially by nations with primitive, teletype- or telephone-based weather communications networks, such as those in the Pacific Ocean, Caribbean Sea, and Central and South America, where communications can be unreliable during hurricane and typhoon season. Hopefully, the new system will have completely replaced the old systems by the turn of the decade. "WAFS is going to completely revolutionize what the world is doing with aviation weather," said Uecker. A number of countries—notably Japan, Germany, and the United Kingdom—have taken the lead in developing advanced, interactive graphical high-altitude weather models and products based on WAFS data. "The weather people in these countries have a mindset for doing this kind of work . . . I just don't see that here yet," said Uecker.

But Uecker has one concern about WAFS. He alleges that in three South American countries (he wouldn't name them but said that they share a common border), local communications agencies have been charging meteorological or aviation authorities exorbitant fees for the "privilege" of accessing the system. Therefore, in some places, WAFS is a clandestine operation.

Uecker believes that in time, the old AFTN would be relieved of its weather distribution burden but would continue to be used as a means of gathering meteorological input from the field into larger and faster circuits.

Better tools

In the earliest days of aviation, pilots had little more available to them for weather information than what was printed in the newspaper. The National Weather Service was primarily focused on providing agriculturally related weather in the form of long-term, seasonal forecasts.

The U.S. Army's Signal Corps provided rudimentary weather information to the military until after World War I, when the various branches of the military created their own aviation weather offices. In the civil sector, the early airlines hired professional meteorologists who gathered and transmitted weather and other operational information by telephone and wireless telegraph networks. In the late 1930s, the U.S. Civil Aviation Agency set up a primitive network of weather offices, which were often colocated with NWS field offices. This network later evolved into today's Flight Service Station system. Only recently have the two agencies untangled their aviation weather relationships.

Traditionally, when pilots began their preflight planning, they gathered up a collection of weather reports and forecasts. For years that required a phone call or a visit to a Flight Service Station or a NWS field office. Meanwhile, the military had dedicated meteorological officers to track weather that could affect military operations, and to some extent, they still do.

The airlines' professional weather briefers stayed on top of changing weather conditions that would affect the carrier's operations, and their knowledge was usually well respected by the pilots who depended solely on company meteorologists and their own weather wisdom to help them decipher flight conditions. In the 1950s, private weather companies that served the air carriers expanded their operations to provide the nascent business aviation segment with reliable weather information. This was the state of the industry until about the early 1980s, when minicomputers, and later, personal computers, began to play a larger role in aviation weather forecasting and dissemination.

For years, weather information was relayed by teletype machine over dedicated circuits. As computers evolved, weather information was formatted so it could be transmitted over existing networks for distribution to smaller computer terminals. Private weather "wholesalers" such as Accu-Weather, Kevouras, WSI, Universal Aviation, and a handful of others reformatted basic NWS weather products, occasionally adding information of their own, and redistributed it to airlines, corporate operators, state and municipal agencies, the military, and the FAA. Considering the level of computer technological development at the time, it was a surprisingly robust system.

The glass teletype

In the early 1980s, a handful of private weather vendors began to experiment with distributing various weather products via computer. Most of the products were lifted right off the NWS weather circuits (these were discussed earlier in this chapter) and were primarily textual in content. However, airline and corporate users wanted more, so the vendors began to provide digital versions of various graphical weather products, such as weather radar, weather depiction charts, forecast prognoses

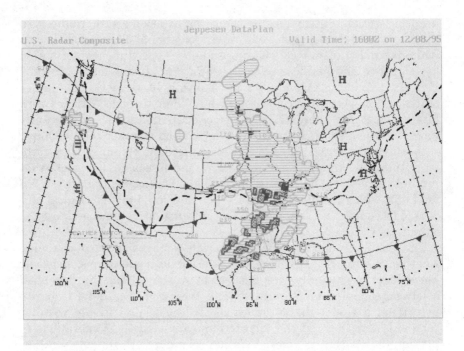

Figure 9.2 Pilots and dispatchers increasingly are turning to commercial weather vendors to obtain flight-planning information. This radar composite chart was obtained from Jeppesen DataPlan via the company's outlet on the CompuServe Information Service. (Jeppesen)

("Prog Charts"), satellite imagery, and so on. The aviation community's response was overwhelmingly clear: They wanted more!

Soon, a cornucopia of weather information could be had via computer, but it was generally too costly for most individual pilots. Pilot organizations such as the Aircraft Owners and Pilots Association (AOPA) and the National Business Aviation Association (NBAA) turned to the government for a solution.

In 1989, after a few years of development, the FAA launched the Direct User Access Terminal (DUAT) system to provide anyone with a personal computer access to government-produced weather products, as well as a few of the products available from the private weather services. DUAT service is provided by Chantilly, Virginia-based GTE DUATS and Turnersville, New Jersey-based Data Transformation Corporation. It was an immediate success, especially with the general aviation community.

Now, anyone with a personal computer (IBM clone or Mac) can simply dial up either DUAT vendor and get all the weather they can handle. The two DUAT vendors provide free weather briefings for anywhere in the United States. Access is via modem directly or with third-party software. Textual weather products include hourly (SA) and amended reports, winds aloft reports, terminal forecasts, and area forecasts. Graphical weather products include radar summaries and weather depiction charts. Character-based weather data is still being transmitted to air carrier cockpits via dedicated very high frequency (VHF) channels operated by ARINC.

Despite this evolution, much weather information has been viewed with a great deal of skepticism or outright mistrust. One of the biggest problems to weather forecasters has been the lack of sufficient data to enable them to predict weather in detailed geographic terms over extended periods of time. Today's weather system provides imprecise information, covers huge areas of geography, and often over- or under-predicts or completely misses areas of adverse weather conditions. Observation sites are too sparsely populated to be meaningful, and forecasts are often based on hours-old data. This lack of data results in overly conservative—even reactionary—weather decision-making by pilots, airline and corporate flight department dispatchers, and ATC and FSS specialists.

Pilots and dispatchers of small airlines and corporate operators on tight turnarounds between flights have a limited time to call FSS, DUAT, or a private weather service for an updated weather briefing. And, as regular users of FSS know, there seems to be an inverse relationship between deteriorating weather and the length of time it takes to get a briefer on the line. NAWPP recommendations call for improvements in the quality of and the ease of access to tailored preflight and inflight weather information.

The ground-side access to weather information seems to be improving rapidly, given the proliferation of electronic weather outlets, both on the Internet and via subscriber services. Development of new ground-to-air architectures to make most, if not all, of the standard weather products available inflight are also ongoing. (These later topics are discussed in greater detail elsewhere in this book).

Cooperative effort

In an effort to improve the weather access situation, a cooperative effort among several federal agencies, including the FAA, the NWS, the NOAA, the DOD, and the National Science Foundation (NSF), has been focusing on applying levels of automation previously unavailable. The result is a combination of tools and technologies that will make weather forecasting more timely, precise, accurate, and accessible.

The name given to this effort is the National Aviation Weather Program Plan (NAWPP), and it has evolved in fits and starts (usually due to budgeting constraints) since its inception in the early 1980s. A number of projects are now being fielded or nearing completion that are likely to advance the science—and art—of aviation weather forecasting. The following section provides a glimpse of some of these projects.

AWIPS. A new high-speed computer workstation and communication network called the Advanced Weather Interactive Processing System (AWIPS) is the centerpiece of the modernization of the NWS. When complete, AWIPS is expected to provide the data processing, data analysis, storage, and display capabilities required to support twenty-first-century forecasting operations at NWS facilities.

AWIPS will be the nerve center of operations at all 118 modernized Weather Forecast Offices (WFO) and 13 regional River Forecast Centers (RFC) in the United States. AWIPS will also be installed at several of the National Centers for Environmental Prediction (NCEP) locations, including the Tropical Prediction Center (TPC), which specializes in tropical weather analysis and forecasts, and the Storm Prediction Center (SPC), which monitors and forecasts conditions that spawn severe thunderstorms and tornadoes.

When completed, AWIPS will be capable of receiving, processing, and helping the forecasters analyze the huge amounts of weather data from the network of Doppler radars (WSR-88D), the next generation of Geostationary Operational Environmental Satellites (GOES), hundreds of new Automated Surface Observing Systems (ASOS), and other data sources such as river gauges and forecast guidance produced at the NCEP, NHC, and SPC. The AWIPS system is composed of two primary elements—the forecast office or components and the communications network.

At the sites where they will be located, the workstations will be the main interface between weather forecasters and the rest of the AWIPS system. NWS forecasters will spend the majority of their time at the workstations interpreting and analyzing data and preparing weather forecast products for transmission. Forecasters will view large amounts of graphic and alphanumeric displays in carrying out the operational mission of the NWS. The AWIPS will store, retrieve, and display a wide variety of hydrometeorological data.

NWS and FAA communications networks will feed data to each AWIPS site, distribute information among the AWIPS sites, and provide for dissemination of information to the public and other outside users. A one-way, point-to-multipoint satellite broadcast service called NOAAPORT will be used to distribute the very large amounts of data products that are collected or produced at NOAA central facilities. Data distributed via NOAAPORT will be accessible not only at all NWS sites but also by any appropriately equipped ground station operated by private sector organizations, universities, etc.

In addition to NOAAPORT, the AWIPS sites will be interconnected by a high-speed data network of terrestrial communications lines. This network will allow two-way, point-to-point communications among the AWIPS sites for the exchange of required data and end products that are locally collected and produced.

Although it is expected the AWIPS will eventually become an important element of the weather forecasting and dissemination landscape, it remains to be seen how quickly it will be available. The DOC's Office of the Inspector General and the GAO have been critical of the AWIPS program, claiming that its development schedule is "risky" and the end product might not meet all the objectives it was designed to accomplish.

AWIPS is the final product of system tests conducted at the NWS Forecast Systems Lab in Boulder, Colorado. AWIPS technical specifications were first written in the early 1980s, so the original specs are considered very old by NWS staffers close to the project. Surprisingly, AWIPS is an open-architecture, integrated system, which is very rare for a government computer system. Its design will accommodate improvements during its lifetime.

AGFS.   The NOAA Forecast Science Laboratory (FSL), based in Boulder, Colorado, collaborates with the FAA and the NWS in the Aviation Weather Development Program. Its stated mission is "to increase the capacity of the nation's airspace while maintaining a high level of safety." The FSL is developing the Aviation Gridded Forecast System (AGFS).

AGFS will provide a high-resolution database of weather conditions, including cloud cover, visibility, winds, and weather radar returns. The AGFS program's primary purpose is to increase the accuracy, timeliness, and spatial resolution of weather products used by air traffic controllers and air traffic managers. More opportunities to develop better weather products now exist because of new observing systems, recent advances in understanding the atmosphere, and better dissemination of meteorological data.

The beauty of this system, when it is implemented, will be the gridded forecast database. The AGFS will be able to provide much more precise and timely weather forecasts for small geographic areas. Currently, many aviation weather products are issued to cover entire states or regions, instead of localized areas, and are thus often overly conservative in their outlook. Developers hope to reduce or eliminate the pilot's least-favorite two words in a weather forecast: "chance of. . . ."

The FSL's Aviation Division is developing the AGFS aviation forecast verification, advanced computing techniques, and aviation weather products for the FAA's Advanced Traffic Management System (ATMS) mentioned elsewhere in this book. FSL's Aviation Division is composed of three branches:

- Aviation Gridded Forecast System
- Product Development and Dissemination
- Model Verification and Production Assistance

The Product Development and Dissemination branch has two groups: the Advanced Traffic Management System and the NOAA Emergency Management Weather Dissemination Project.

To complete the link to the pilot, the FAA is developing the Aviation Weather Products Generator (AWPG). This system will draw information from the AGFS to provide high-resolution analyses and "nowcasts" (current weather) tailored to users' needs. However, due the complexity of AGFS and AWPG, both will be costly to complete.

Indeed, this program, like many other FAA efforts, has been in development for more than 15 years and appears to be gridlocked in the definitions/requirements/specifications/funding/R&D loop. Unfortunately, by the time the AGFS program does get the necessary funding and approvals, computer technology will have superseded it, so the program will be viewed as an outdated answer to an ongoing need, and the process will start all over again.

AWOS/ASOS. Using automated sensors, the Automated Weather Observing Station (AWOS) will observe, format, archive, and transmit observations—you guessed it—automatically. After gathering information, the system disseminates the information to pilots via a computer-generated voice communications system. Information is also routed to the NADIN circuit for delivery to standard aviation weather outlets, such as DUAT. If you live near an airport with an AWOS, you can call it up to obtain current weather information for that airport. When preselected weather element thresholds are exceeded (e.g., the visibility decreases to less than 3 miles), a "special" report is transmitted.

The FAA bought 200 AWOSs for airports that do not have control towers or human weather observers. AWOS is available commercially and was purchased to meet the immediate need for automated weather during the development of the more sophisticated Automated Surface Observing System (ASOS). In addition to gathering the weather information obtained by AWOS, ASOS will be able to identify types and amounts of precipitation and provide displays for use in control towers.

When installation is completed, which is expected in the late 1990s, the ASOS sites will serve as the nation's primary surface weather-observing network. ASOS will significantly expand the information available to forecasters and the aviation community. The ASOS net-

Figure 9.3 A NASA demonstration program called Cockpit Weather Information Network (CWIN) enabled pilots, using a touch-screen interface, to call up radar summaries, surface observations, terminal forecasts, ceiling and visibility, and other data, as well as respond to alphanumeric ATC clearances supplemented by synthetic voice messages. Such a system, NASA developers estimate, would cost about $2000 per aircraft. (NASA)

work will more than double the number of full-time surface weather-observing locations. A basic strength of ASOS is that critical aviation weather parameters are measured where they are needed most: airport runway touchdown zones. ASOS detects significant changes, disseminating hourly and special observations via the networks.

The ASOS program has been delayed due to installation incompatibilities with control tower displays, the development of freezing rain sensors, and the need to obtain additional funding required for additional ASOS units.

ATMS. The Advanced Traffic Management System (ATMS) is a Forecast Science Laboratory (FSL) joint effort with the Volpe National Transportation Systems Center (NTSC) and FAA. ATMS develops and demonstrates the use of advanced high-resolution meteorological data sets for air traffic management. These data sets include high-resolution radar data, lightning strike data, text information, and the AGFS. Additionally, FSL develops weather displays for air traffic managers that are compatible with their air traffic data. Also, the AGFS wind data is integrated into aircraft trajectory algorithms that predict aircraft movement/locations to make airspace congestion predictions. Both of these prototype capabilities are integrated into the Aircraft Situation Display (ASD), which is used by FAA traffic managers for strategic planning of the airspace system.

ITWS.   Airport tower controllers rely on a variety of sources of weather. Interpretation of weather data is a manual, often time-consuming process. Also, current weather data-collection systems don't easily provide the ability to create short-term (0 to 30 minutes hence) forecasts, or "now-casts."

A system called the Integrated Terminal Weather System (ITWS) will soon start to change this shortcoming. ITWS will automatically gather and integrate data from a variety of sources, including Terminal Doppler Weather Radar (TDWR), low-level windshear (LLWS) detection systems, and aircraft in flight to provide short-term automated weather and predictions in easily understood graphical and textual forms to air traffic controllers. ITWS's products will include windshear and microburst predictions, storm cell and lightning information, terminal area winds aloft, runway winds, and short-term ceiling and visibility predictions. ITWS has been successfully demonstrated at several airports and 37 systems are expected to be installed by 2001.

Military Weather Systems

The U.S. Air Force, Army, Navy, and Marine Corps play an important role in gathering weather information and producing weather products for their own use, as well as for the civil sector. The following sections briefly describe the armed forces' contributions and recent computer technology improvements.

Air Force

The Air Force's mission includes providing aviation weather services to active and reserve Air Force and Army units, National Guard units, and other selected DOD agencies. This mission involves other centralized and combat support roles. The Air Force produces a wide variety of products and specialized services. These can be provided by Base Weather Stations (BWS), Weather Support Units (WSUs), or staff meteorologists (STAFFMETs).

The Air Weather Service (AWS) oversees transition of new technologies to operational capabilities; manages Air Force-wide weather systems; and provides standardization, interoperability, and evaluation of the U.S. Air Force Weather Support System. AWS also provides centralized weather, climatological, and space environment support to designated users through three centralized support organizations: Air Force Global Weather Central (AFGWC) at Offutt Air Force Base (AFB) in Nebraska; U.S. Air Force Environmental Technical Applications Center (USAFETAC) at Scott AFB in Illinois; and Air Force Space Forecast Center (AFSFC) at Falcon AFB in Colorado.

The AFGWC supports the National Command Authority, the National Military Command System, and the National Security Agency with tailored environmental support products disseminated to these users worldwide. AFGWC uses mainframe production computers, a cluster of super minicomputers, and an interactive graphics and imagery system to run worldwide analysis and forecast models to define the current and future states of the atmosphere. High-resolution forecast models enable AFGWC to accurately focus on remote areas worldwide. AFGWC is also the backup for the NWS facsimile network and NWS National Severe Storms Forecast Center (NSSFC).

USAFETAC provides military climatological support to the Air Force, Army and other government agencies. AFSFC provides a dedicated space environmental support capability for communications, surveillance, and space operations. Disturbances in space (such as solar storms) can affect aviation through loss of high-frequency communications and impacts to the Global Positioning Satellite (GPS) system.

Regional weather centers support military requirements in specific areas, such as Europe, Korea, and Panama. Combat theater support is provided by theater forecast units and deployed weather teams (WETMs). They generate tailored weather products and mission forecasts. The Air Force communications system consists of conventional weather teletype (TTY) networks, high-speed automated digital systems, long-haul and point-to-point data circuits, fax networks, and other systems.

The Automated Weather Network (AWN) is the backbone of military weather communications used for collecting and distributing alphanumeric, binary, and graphical weather data. High-speed computers deliver foreign and domestic weather data to designated users. The AWN also interfaces with the Navy Fleet Oceanographic Center, in Monterey, California, and the NOAA National Meteorological Center (NMC). Overseas alphanumeric data collection and dissemination networks are driven by the AWN Automated Digital Weather Switch (ADWS) computers at Hickam AFB in Hawaii and Royal Air Force Base in Croughton in the United Kingdom. In the United States, the AWN communications hub is located at Carswell AFB in Texas.

The Communications Front End Processor (CFEP) at the AFGWC is the hub of the transmission system, providing gridded data, vector graphic data, fax, and TTY outputs to worldwide military users. The CFEP drives the separate networks serving the continental United

States (CONUS), Alaska, Europe, Central America, and the Pacific, using computers to store and forward required products.

The computer-based operations of AFGWC uses a "build-and-apply" concept. AFGWC "builds" a real-time, integrated environmental database by using worldwide weather data, relayed to AFGWC by AWN and blended with civil and military weather satellite data. Computer algorithms are used to process the data to construct models of the atmosphere and to forecast its future behavior. AFGWC then "applies" the model output to generate tailored products designed to meet the specific user mission needs. AFGWC forecasters use the Satellite Data Handling System (SDHS), a unique human-machine interactive computer system, to display and assimilate conventional and satellite data in graphical form to aid them in preparing a multitude of products.

Recent Air Force aviation weather support system enhancements include

- *AFGWC Software Improvements Program (ASIP).* The ASIP will improve and standardize documentation of the approximately 1,165,000 lines of code in current use by the AFGWC Weather Infor-

Figure 9.4   The CWIN terminal was tested aboard a NASA Boeing 737 fitted with an extra, experimental cockpit in the fuselage. A standard cockpit is where it should be. (NASA)

mation Processing System at Carswell AFB, Hickam AFB, and RAF Croughton.

- *Advanced Computer Flight Plan (ACFP)*. The ACFP will replace a leased, optimized computer flight plan system and an existing Air Force-owned flight simulation model.

- *Satellite Data Handling System (SDHS II)*. SDHS II is a life-cycle replacement and upgrade system of AFGWC's interactive workstations that support worldwide Air Force weather units.

The system has more than 70 computers, 38 forecaster consoles, a database subsystem, and communications interfaces.

Additionally, the Air Force is upgrading a variety of automated weather systems. Among these are

- *AWDS and Transportable AWDS (TAWDS)*. This modular system automates the handling of weather data by incorporating the latest state-of-the-art data processing, communications, and display technologies with forecast techniques. TAWDS supplies in-theater support.

- *Combat Weather System (CWS)*. CWS will combine computer software and hardware, weather sensors, and communications hardware to provide a weather support capability for future combat operations. CWS will be a highly automated, small, lightweight, and tactically deployable system.

- *Fixed Weather Observing System (FWOS)*. This system will replace and upgrade existing weather sensors. FWOS might use sensor technology developed for the Combat Weather System (CWS) and the NWS's Automated Surface Observation System (ASOS) program.

Army

The U.S. Army has existing systems to observe both upper-air and surface weather and to process weather satellite imagery. The Army's upper-air system is the AN/TMQ-31. It is used by artillery meteorologists to obtain information over a battlefield area. The upper-air observations are also forwarded to U.S. Air Force WETMs to support aviation missions.

Navy and Marine Corps

Within the U.S. Navy, meteorological and oceanographic support is provided by the Naval Oceanographic and Meteorological Support System (NOMSS). Primary support for the NOMSS is provided by field activities, detachments, and units assigned to the U.S. Navy Naval Oceanography Command (NAVOCEANCOM). Shore activities within NAVOCEANCOM having meteorological responsibilities include a primary numerical processing center, three regional Naval Oceanography Command Facilities (NAVOCEANCOMFACs), and nearly 50 Naval Oceanography Command Detachments (NAVOCEANCOMSETs).

The Fleet Numerical Oceanography Center (FLENUMOCEANCEN), in Monterey, California, is the master computer center for NOMSS. FLENUMOCEANCEN receives global data to generate numerical weather analysis and prediction products and applications products for distribution to NAVOCEANCOM users and for use by NOMSS activities to generate meteorological support products and services. The FLENUMOCEANCEN is the hub of the Naval Environmental Data Network (NEDN) that links the major NAVOCEANCOM activities.

Naval and Marine Corps aviation weather products are provided directly to the user through a host of intrabase communications networks or across the counter.

Among the newer naval aviation weather systems are

- The Aviation Support Display System (ASDS). The NAVOCEANCOM supports more than 50 Navy and 11 Marine Corps air stations by providing flight forecasts, local area forecasting, and climatological support services. Until recently, these services were largely manual. The ASDS program provides 67 computerized systems that can receive, store, enhance, annotate, and display data at the weather forecaster's workstation. The incoming data includes alphanumeric, color graphic, radar, lightning, and satellite information. ASDS interfaces with the Naval Oceanographic Data Distribution and Expansion Systems; Air Force AWN; digital facsimile broadcasts (DIFAX); and Geostationary Operational Environmental Satellite (GOES).

- Tactical Environmental Support System (TESS). TESS Version 3.0 is a modular, computer-based support system designed to provide Navy decision-makers with meteorological and oceanographic assessments and forecasts and to integrate air and ocean data with sensor/weapon platform parameters to assess system performance. TESS 3 functions as the operational, resident air/ocean master database.

Nonfederal weather systems

Nonfederal systems consist primarily of the systems operated by individual states and those operated by commercial weather services. Both of these resources are playing an increasingly significant role in providing weather information to the aviation community.

Although it is discussed at length in Chapter 2, it should be noted here that the Internet is quickly becoming another major resource for aviation weather as providers and pilots are finding new methods to deliver and receive aviation weather products. Private weather services are creating increasingly useful and effective means of providing pilots with a growing array of graphical weather products. Although the Internet medium is still maturing, it suggests a whole new method of exchange for providers and users alike.

Storming the cockpit

The next step in aviation weather technology, more than likely, is going to involve data link technology and, later, advanced, interactive cockpit displays. Some of these systems, in development since the 1980s (and described in Chapter 2), will include real-time, computer-generated graphical or alphanumerical information about hazardous weather such as thunderstorms, turbulence, icing, and windshear. A variety of techniques have been developed to make this possible.

Other possibilities include a system being developed by the NWS's Severe Storms Forecast Lab in Boulder, Colorado, that could enable pilots to call up computer-generated, full-motion, color-coded, three-dimensional views of convective activity and windshear. The system would provide flight crews with a remarkably vivid image of hazardous weather conditions in a terminal area, for example. An arriving jet could easily determine whether or not it would be safe to commence an approach into an airport in the vicinity of a thunderstorm.

FAA's Administrative Networks and Data-Processing Systems

The FAA employs a wide area network (WAN) to link its regional offices and Air Route Traffic Control Centers (ARTCCs) with its Washington, D.C., headquarters. The system is called ADTN-2000 (Administrative Data Telecommunications Network).

ADTN-2000 is a completely outsourced, turn-key system leased from and operated by GSI, Inc. The contract for this system was awarded in early 1995, and it replaced an earlier, FAA-owned ADTN system. The new system provides an upgraded telephone service and other features. The system is designed to provide the agency with a modern administrative data and telecommunications structure, and leasing the system enables the agency to keep up with the rapid pace of development in telecommunications.

ADTN-2000 consists of Cisco routers and provides switched X.25 telecom service. The centers and regions are connected via local area networks (LANs), which means the system is able to link 33,000 desktop computers. It also provides dial-up connectivity to the FAA's mainframe computer services.

The FAA's Airmen and Aircraft databases are maintained on outsourced mainframe computers operated by EDS, Inc., which is the prime contractor. That contract was awarded in January 1992 under the FAA's CORN (Computer Resource Nucleus) effort to modernize its computer systems. The system currently uses an Amdahl mainframe computer. The IBM-based database software is provided by Software AG. The computers are currently located in EDS's corporate headquarters in Plano, Texas, and the FAA also has a fully redundant backup site in Flint, Michigan. EDS is responsible for meeting the FAA's capacity demands over the life of the contract.

Airmen information system

If you're a pilot, licensed dispatcher, or maintenance technician, you might have wondered where your certificate comes from. The answer is Oklahoma City (OKC), and the system that produces it is called CAIS. The Comprehensive Airmen Information System was designed to support the agency's airmen certification program and to issue permanent airmen certificates. CAIS maintains data identifying U.S. certified airmen, certificates they hold, and the ratings and limitations on their certificates. The system provides information to FAA field offices, other federal agencies, state and local government agencies, foreign governments, and the general public regarding the certification of airmen. Personal airman information contained in CAIS is protected by the Privacy Act, which restricts its release. CAIS is the source for statistical data providing information regarding the numbers and types of FAA certified airmen.

CAIS is an operational system supporting major functions performed by the Airmen Certification Branch, which are conducted through interaction with the system. The system maintains databases of information regarding the airman's physical description, types of certificates held, certificate number, ratings and limitations, date of issue, record location, and the airman's medical certificate's date, class, and limitations.

Aircraft registration system

Similarly, information about U.S. civil-registered aircraft are also maintained on a mainframe computer system in "Oke City." The Aircraft Registration (AR) System was designed to support the registration of U.S. civil aircraft and the issuance of the certificates of aircraft registration. The AR automated system maintains the data regarding the identification of U.S.-registered aircraft, information about who holds the certificate of registration, and certain airworthiness data. The assignment and reservation of U.S. registration marks is controlled through the AR system. The system carries the description of each aircraft registered in the United States, including aircraft make, model, serial number, along with registration information, including the registered owner's name and address.

The system serves as the United States Civil Aircraft Register as required by the International Civil Aviation Organization (ICAO) and provides information to foreign governments and the general public regarding the U.S. civil aircraft fleet. The system provides national online access to FAA field offices through the Aviation Safety Analysis System (ASAS), and to the National Aircraft Registry Information subsystem (NARI). The AR system is a source for statistical data regarding the numbers and types of aircraft registered in the United States.

The AR system supports major functions performed by the Aircraft Registration Branch that are conducted through interaction with the system. The system provides precise information regarding record location and availability, certificate issuance, notice of renewals, document receipt, processing, tracking, and updating.

The Aircraft Document Index System (AD), which is a subsystem of AR, supports the aircraft registration system by collecting data regarding receipt of documents. It identifies the parties and collateral covered by the documents received and, in the process, identifies the aircraft records needed for review. The document index is used by the aviation community to verify official filing of documents. After reviewing an aircraft record, the index is checked to validate that no other filed documents affecting the aircraft have not yet been associated with the aircraft record. The system services not only the FAA, but also the aviation public. Inquiries can be made by the public to obtain information directly from the system through terminals made available in the public documents room and at inquiry telephone stations within the Aircraft Registration Branch.

Flight inspection fleet management

In 1995, the FAA selected St. Louis, Missouri-based INAIR, Inc.'s fleet management system to help the agency keep track of its pilots, who fly more than 50 of its flight inspection and transport aircraft. These aircraft are used to verify the accuracy of the nation's navigation and communications facilities. Each day, the pilots of the FAA flight inspection aircraft (typically a Beech King Air) fly extensive, grid-pattern flight tracks while technicians in the back of the airplane check the reception, signal quality, and accuracy of facilities in the area and coordinate with technicians on the ground to calibrate the systems. Often, when there's been an accident or major incident, the flight inspection team will go to the accident area to verify any navaids or other communications facilities that might have been in use at the time of the accident. These aircraft are also used in support of the FAA's many field offices.

Under pressure to comply with its own regulations, the FAA began a two-year search for an off-the-shelf, integrated crew scheduling and management system before choosing the INAIR package. INAIR, which evolved from a defense-industry software firm about five years ago, initially developed its modular operations control system for regional and business aviation users. INAIR now counts among its clients several regional passenger and heavy-transport freight carriers and at least one Fortune 100 operator. The system selected by the FAA is initially a beta site PC-based version of INAIR's current LAN product, but agency and INAIR programmers are in the process of converting the system to a WAN client-server system, a process that will take about 12 months to complete, according to FAA project manager Travis Ray. The final configuration will link the agency's flight operations LANs in Oklahoma City and Herndon, Virginia, as well as other regional sites.

The new system, when complete, will enable the FAA's flight inspection team to function more like a regional airline operator. It provides

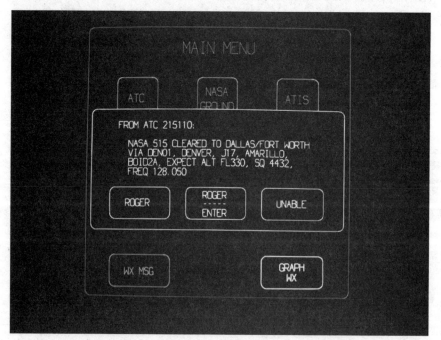

Figure 9.5 One CWIN display featured a module for data-linked ATC messaging. The pilots could respond to clearances via the system's touch-screen display. (NASA)

tracking of crew duty hours, medical certificates, and recency of experience and training records.

The INAIR selection will enable the agency to eventually replace the system it currently uses, an OKC-based mainframe computer system that was developed in-house in 1977 mainly to provide maintenance inventory support functions. Making changes or requesting reports from that system can take a week or more, and the branch wants a little more flexibility and better response.

INAIR officials say the company is in a codevelopment relationship with Sabre Decision Technologies to create other modules for shop floor and engine shop tracking. If the government decides to mandate that the FAA manage all of its nonmilitary aircraft, INAIR stands to be well-positioned to expand its operator base significantly. The company also makes a fleet tracking system.

In addition to the INAIR system, the Oklahoma-based branch uses the Aviation Standards Information System (ASIS) for flight scheduling and dispatching. ASIS was developed in-house. The system's ability to provide centralized scheduling will improve the agency's flight inspection area offices to better coordinate their operations, as well as allow better fleet parts inventory and maintenance tracking at the agency's OKC facility. Ultimately, as many as 1000 agency employees will use the system.

Another system, provided by the Gull Aviation division of Parker-Hannifin Corp., is used to schedule flight inspection tasks efficiently. The OKC Flight Inspection Central Operations facility has an Aircraft Situation Display (ASD) system, so if a flight inspection aircraft is on an IFR flight plan it will show up on the ASD. The branch uses another system called Airnav to maintain its database of navaid latitudes and longitudes and other airways facility information.

"Of course, it's not unusual for our aircraft to file to one location and wind up somewhere else," said Hal Walls, manager of technical support for the flight inspection facility. "That's just the nature of our work." The branch is investigating the possibility of using satcom and data link to keep in touch with its aircraft, but it has also found that a simple, commercially available satellite-based paging system is also effective.

Walls said he's happy about the requirement to operate more like an air carrier. "It pretty much guarantees that our pilots will get the training they need," he said. "We used to have to beg for it."

Reaching out: Electronic bulletin boards and the Internet

In an effort to broaden its reach to the aviation community the FAA began implementing electronic bulletin board system (BBS) access in the early 1990s and jumped onto the Internet in mid-1995. The two points of access have been phenomenally successful for the agency.

"Although the BBS activities are under my organization," said Mike McCann, who heads up the FAA's Office of Information Technology, "we really haven't paid much attention to them—they're really self-sustaining. Our corporate BBS is being used about 75 times a day. We get users from both inside the FAA and the aviation community as well.

"We started an initiative to shut down the BBSs in favor of the Internet, but so far we're getting about 10 to one against doing that, so we'll keep running it as long as we need it. And speaking of the Internet—I've got some interesting statistics on that, too. We have one main and two other significant sites that are hosting home pages—and that incorporates all the major organizations—air traffic facilities, airports, regulation and certification, as well as the two major FAA centers (the Mike Monroney Aeronautical Center in Oklahoma City and the Technical Center in Atlantic City). The FAA home page is getting about 150,000 hits per week. When we first brought it up about nine months ago [June 1995], 5000 hits a week was a really exciting milestone to reach. I was just flabbergasted when we got up into the 100,000-hits-per-week range."

I asked McCann if the agency would be expanding its Internet presence.

"The content is expanding significantly," he responded. "And now the agency is also building an intranet capability, which will allow us to facilitate our intraagency communications. It will do things that our bulletin boards do not provide. The ADTN-2000 will be the 'pipe' that carries this intranet. Many corporations have intranets now, like Boeing, GM, and others. They're all firewalled systems, except for those items where you want people on the outside to have access. Even GSA [Government Services Administration] has recently announced their Advantage system for purchasing—that's the old GSA Schedule concept also on the Internet. We'll be connecting directly to that because that will be our preferred means of purchasing information technology-related services in the future."

We took another look at the FAA's Internet presence in Chapter 2.

The following are descriptions, Telnet URLs (where known), and telephone numbers of the current FAA-sponsored BBSs. These numbers have been verified as of June 30, 1995:

- FAA Corporate Bulletin Board, (800) 224-6287. Sponsored by the Office of Information Technology (OIT), this board carries information on a number of agency programs as well as general information. This board also provides a dial-out gateway facility to other FAA bulletin board systems. Choose "Interchange—Gateway to Other FAA BBSes" to use this facility. The Telnet URL is telnet://faacbbs.hq.faa.gov. (Note: There is no download available when using Telnet to access this BBS.

You must have a Telnet application configured in your browser for this link to work.)

- Airports Bulletin Board, (202) 267-5205. This board carries information of interest to airport operators and designers.

- Air Traffic Operations Bulletin Board, (202) 267-5331. This board carries polls, questionnaires, and classified ads of interest to the air traffic operations community.

- Air Transport Division Bulletin Board, (202) 267-5231. This board supports the following special interest areas: transport category aircraft, Flight Standardization Board (FSB) documents, draft documents, and comments on draft documents. It also contains a forum on flight safety.

- Environment and Energy, (202) 267-9647. This bulletin board supports conferences on noise models and economic costing models.

- Aviation Rulemaking, (202) 267-5948. Sponsored by the Aviation Rulemaking Advisory Committee (ARAC), this board contains a list of future meetings and information on special interest groups.

- Portland MMEL BBS, (207) 780-3297. This bulletin board contains selected FAA information publications of interest to the aviation community, as well as information about the status of various MMELs.

- Safety Data Exchange BBS, (800) 426-3814. Operated out of Kansas City, Missouri, this board contains accident report forms and information on general aviation safety issues.

- Aeromedical Forum BBS, (202) 366-7922. This board seeks to promote the exchange of information about aviation medicine. It hosts conferences of interest to regional flight surgeons, not all of which are open to the public without prior approval, which can be requested through the BBS.

10

Data Link:
Tomorrow's Digital Airways

It should be evident by this point that aircraft are quickly becoming airborne digital networks. At the same time, ground-side computer-based technologies promise to leverage the efficiency of aircraft and the national airspace system while increasing safety. The next step toward creating a seamless, digital aviation environment is the implementation of data link communications technology—a digital airway environment, if you will. Data link will finally connect the ground side with the aircraft in real time.

Data link, say its proponents, is going to become the cornerstone of much of aviation's future. By introducing a new way to communicate, data link will provide significant benefits to all airspace system users. Proponents of data link technology—primarily the FAA, ICAO, and the OEMs, say that data link implementation, over time, coupled with advances in functionality, will lead to increased safety, efficiency, and capacity. These improvements in airspace management will contribute to increased savings in aircraft operational costs, savings that are estimated in the millions of dollars.

At this point, there don't appear to be any groups that really oppose data link or doubt that data link will be implemented, but some believe the devil may be in the details. There are those who question *how* data link will ultimately be implemented, what it will cost, and how soon it really can become a part of the national and international airspace infrastructure. The current Free Flight debate revolves around some of the implementation issues that go hand-in-hand with data link.

Nevertheless, I think it's a safe bet that, ultimately, some form of data link will be implemented. The airlines, general aviation, air traffic management, and even passengers will all benefit from the modernization that data link will enable. Regardless of your relationship to data link, becoming familiar with concepts like ADS, ATN, FANG, FANS, CNS/ATM, and others will help you to understand some of the issues it raises and, more importantly, the potential benefits and opportunities of data link.

As we saw in Chapter 2, today's advanced avionics are becoming the cockpit workstations on the new communications, navigation, and surveillance/air traffic management (CNS/ATM) environment that is emerging from various industry initiatives to improve the safety and efficiency of the world's airspace.

The CNS/ATM concept includes the communication systems used by aircraft, ATC, and terrestrial links to flight operations, such as an airline flight dispatch department. It can include any of the data link concepts we're about to discuss. Navigation refers to all systems used by an aircraft to fix its position in space and fly from one point to another. These systems can include the Global Positioning Satellite (GPS) system, the VOR Airway system, and, to a lesser extent, Loran-C, NDB, VLF/Omega, and onboard Inertial Navigation System (INS) systems. Surveillance includes ATC surveillance radar and automatic dependent surveillance (ADS), which relies primarily on satcom to function, especially in oceanic areas.

To understand data link in an aviation computing systems context requires an understanding of satcom, FANS, and ADS, and the initiatives to implement them regionally and globally. Satcom was discussed

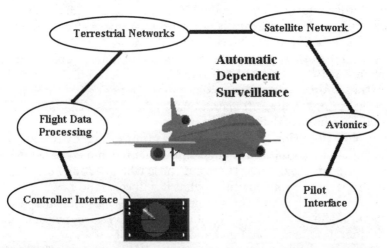

Figure 10.1   A fundamental element of the data link-based CNS/ATM effort is automatic dependent surveillance (ADS), which will link pilots, controllers, and airline operations. (Author)

at length in Chapter 2, so we'll start here with FANS. Although it is currently being implemented in the Pacific Ocean arena, it is likely to become a model for subsequent areas of the world.

FANS

One of the most significant CNS/ATM initiatives in a generation is FANS (Future Air Navigation System). This international effort is likely to have far-reaching implications for aviation computing systems, especially avionics, air traffic management, and airline operations. It relies heavily on data link and satellite-based navigation technology—and a lot of international cooperation—to make it happen.

FANS is a term describing a whole set of concepts for air traffic management tools and avionics systems of the twenty-first century that will be playing an increasingly important role in information processing and telecommunications. Some of these systems are already in place, others are awaiting implementation funding, and others are still in development.

The FANS concept was developed by the International Civil Aviation Organization (ICAO). ICAO is a specialized agency of the United Nations made up of 183 member states organized to foster the development of international air transport by developing international standards and procedures to promote safety, uniformity, and efficiency in air navigation. ICAO publishes a variety of technical, economic, and legal documents, and its members meet frequently.

FANS, the latest ICAO initiative, was adopted by ICAO's member countries in September 1991. FANS proposes the development of a global navigation satellite system (GNSS) and other initiatives to improve navigation and communication, particularly in parts of the world that have inadequate, outmoded, or nonexistent facilities.

Until ADS, which includes satellite-based data link-based position reporting and ATC clearance communications, becomes widespread, controllers managing air traffic navigation in transoceanic or remote areas must rely on HF radio position reporting. HF techniques were developed prior to World War II, when Pan Am's amphibious China Clippers proved the first transoceanic routes to Asia, South America, Africa, and Europe.

Although the navigation systems installed in today's aircraft are very accurate and reliable, current ATC procedures for operations in remote areas of the world (60 percent of the world is beyond the reach of ATC radars) still require manual data entries of aircraft positions and waypoint insertion in the aircraft's FMS—with the possibility of crew error. (An incorrect manual position entry was presumed to have led to the downing of Korean Air Lines Flight 007 over far-eastern Russia on September 1, 1983, by a Soviet fighter after the 747 flew through Soviet airspace near Sakhalin Island; 269 people were killed.)

It falls to the controllers—with no automated backup systems—to make sure their assigned aircraft stay on the their assigned flight plans. To make this possible, traffic management procedures require that aircraft be separated by hundreds of miles to avoid any possibility of collision. Obviously, this separation requirement puts a severe limit on the amount of traffic that can be taken over well-traveled routes such as New York/London, San Francisco/Tokyo, or Melbourne/Tokyo.

The first area to see FANS implemented is a section of the Pacific Ocean region. This area is considered ideal for trying out the FANS concept because it is low in traffic, politically friendly, and the CAAs in the area are eager to improve their ATC infrastructures.

This first FANS implementation effort is known as FANS-1 and is being fostered by the FAA in the United States, the CAA of Australia, the Airways Corporation of New Zealand, plus CAA Fiji, Boeing, Honeywell, United, Quantas, Cathay Pacific, and Air New Zealand air lines.

FANS-1 specifies an initial implementation based on FANS-1 data link communications using the ARINC 622 protocol, Inmarsat satcoms, and ATC workstations capable of integrating ADS, pilot/controller data link, and ground-to-ground communications. The aviation agencies involved have stated their intentions to commit the funding to set up the basic infrastructure in Australia, New Zealand, and U.S. West Coast ATC facilities responsible for handling transpacific traffic (primarily Oakland Center). Unfortunately, at this writing, only New Zealand has installed the full complement of ATC hardware and software, and only partial installations have been accomplished elsewhere.

Among the key concepts to FANS-1 implementation is Oceanic Data Link (ODL), which is an element of the FAA's Oceanic Automation Program (OAP). ODL makes use of satellite communications-based controller-pilot data link communications (CPDLC). ODL was initially developed for the Airways Corporation of New Zealand by CAe Electronics of Montreal in an installation called Oceanic Control System. The new system allows controllers to track aircraft via regular satellite data link updates of an aircraft's position on a visual situation display. New Zealand became the first country to complete testing of the system in September 1995. The tests were intended to finalize ODL procedures before the system was to be installed in the FAA's Oakland Center facility. Testing involved the "shadowing" of aircraft equipped with satcom data link systems supplied by Honeywell Commercial Avionics of some 900 revenue flights that used HF radios for routine position reporting. The FAA's Tech Center in Atlantic City, New Jersey, reviewed the test data.

Data link benefits

Now that we have established the international nature of FANS, we can discuss how data link will likely become the *lingua franca* of FANS. The use of data link offers some real benefits. By giving the aircraft a new way to communicate, data link will likely provide airline flight crews and aircraft operators access to more information, while increasing the efficiency and reliability of current communications. Proponents assert that data link's benefits include

- More direct routes
- Less vectoring
- Fewer holds
- Reduced ground delays
- Reduced communications errors
- Reduced flight times

To the GA pilot, data link might be the key to receiving affordable services in the cockpit. By transforming the present communications system, data link promises to provide the GA community with increased functionality and savings. Benefits to GA could include

- Increased utility
- Improved situational awareness
- Better routes and altitudes
- Improved safety
- Reduced communication congestion and delays
- Improved cockpit information management
- Increased operational opportunities

Data link could provide significant benefit to advanced air traffic management and greatly improve ATC operations. By increasing a controller's ability to communicate in a timely, reliable, efficient manner, data link enables the controller to meet the FAA's ATC mission: "The safe, orderly, and expeditious movement of traffic through the National Airspace System." Data link provides ATM with

- Increased efficiency
- Better workload management
- Increased sector throughput

Figure 10.2 AlliedSignal's Global Wulfsberg GNS XLS flight management system is typical in that it provides the flight crew with keyboard access to the aircraft's navigation and performance databases and flight computers, as well as the aircraft's nav sensors. Teaching pilots how to use such systems is a quickly growing training field. Some day, such "data terminals" will be linked to ground-based computers. (AlliedSignal)

- Reduced restrictions
- Reduced voice congestion

Although passengers cannot see the visible impact of data link, they might experience marked improvements in aircraft operations that translate into benefits for the airline passenger. Reduced flight times and enhancements to safety will enable passengers to utilize the air transportation system with the confidence of knowing they are using the safest and most efficient transportation system in the world.

Data link technology will provide aircraft operators with tangible operational and financial benefits. With data link, aircraft can be rerouted after departure to take advantage of wind-optimized routings, reducing fuel requirements and increasing payloads. Eventually, separation of traffic over oceanic and remote regions will be reduced, which will increase airspace capacity to accommodate growing traffic levels.

How data link might work

Although many details still must be ironed out, field trials of some data link applications have made it appear that it is at least possible. But before we can discuss some of data link's applications, we'll need a better idea of how data link will work.

Telecommunications is changing our world. Data link extends telecom to the world of aviation. Data link is a telecom network that will allow for digital transmission of data to all users within the national airspace and, ultimately, the entire global airspace.

The digital process converts information into zeros and ones by automation systems and sends it to its destination via one of several "data links." The receiving automation systems convert the digital data back into an understandable format, depending on the nature of the information. Data link can advance aviation communication by providing a means to send more information with greater efficiency, reliability, and ease.

By introducing digital communications into the current voice-only airspace management environment, data link will change the way

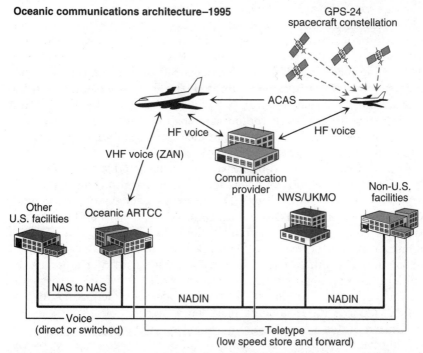

Figure 10.3 The Future Air Navigation System (FANS) concept calls for a variety of communications media for transmitting communications, navigation, and surveillance (CNS) data, including satcom, HF radio, automatic dependent surveillance–broadcast, and traditional terrestrial telecom networks. (FAA)

pilots, controllers, and airspace managers transmit and receive routine information. No longer will pilots and controllers be restricted by the limitations of the voice-only environment.

The data link system backbone includes multiple independent subnetworks:

- VHF radio
- Mode S secondary surveillance radar
- Satellites

The Aeronautical Fixed Telecommunications Network (AFTN), acts as a "glue" binding these subnetworks with automation systems to form a worldwide interoperable data communications network. Depending on the requirements, a data link subnetwork can send data using either addressed or broadcast methods, thus allowing for greater overall system functionality. Addressed messages are sent point-to-point to a specific user anywhere in the world, while broadcast messages are sent to any user within the broadcast area vicinity. The primary means for data link communications that are under consideration are via VHF radio, Mode S, and satcom.

Voice communication in the VHF band is used for aeronautical communications worldwide. Almost all civil aircraft are equipped with at least one VHF radio, which they use as the primary means for transmitting and receiving ATC and airline operations information. Because this voice system is inherently limited, the FAA is working with the international aviation community to define standards for VHF digital radio. Future upgrades in radio technology to all-digital systems can provide a cost-effective data communications capability to all users within the VHF ground station line of sight.

Mode S secondary surveillance radar is designed with an integral data link that meets many needs for air-to-air, ground-to-ground, and airport surface communications. This beacon sensor system will replace ATC beacon sensors (ATCRBS—pronounced "at-crabs") and provide an added capability: digital communications. Mode S will provide data communications services to all equipped users within Mode S ground sensor line of sight. ATCRBS is currently used at airports across the United States and many other parts of the world, providing airport surface surveillance data to air traffic management facilities and controllers. Mode S radar is being installed at nearly 150 airports across the United States, providing an upgraded surveillance capability to the existing radar system. With planned national implementation by the end of 1996, Mode S radar will also provide digital communications capability for all Mode S equipped aircraft.

Satcom has dramatically changed when, where, and how the world communicates. Satcom provides rapid, accurate communications almost anywhere in the world, linking ground and airborne users via space-based satellites. This capability is being demonstrated in oceanic airspace for controller/pilot communications and aircraft position reporting. Satcom will also be used by the United States to broadcast data to augment the integrity and accuracy of GPS in the GPS Wide Area Augmentation System (WAAS).

ATN

The Aeronautical Telecommunications Network will connect a diverse set of air-to-ground and ground-to-ground communications systems, such as VHF, Mode S, and Satcom, to provide global information transfer between air traffic management, aircraft operations, service providers, and passengers.

ATN routers are the foundation of the ATN's interoperable communications system, providing a "connection" for seamless data link communications worldwide. ATN routers are computers that will keep track of aircraft locations and use the best available data link to communicate between the aircraft and the ground networks.

The requirements for the ATN are being developed within the international community to satisfy the complex telecommunications needs of countries around the globe. To foster the development and adoption of ATN technology within the United States, the FAA has formed the ATN Consortium.

A consortium of 11 U.S. airlines has formed a private, for-profit company called ATN Systems, Inc. This consortium works in partnership with the FAA, research organizations, and industry on the development and implementation of ATN routers in the United States. The consortium is off to a fast start, having released a request for information a week after the consortium was formed—something that would take a federal agency months to do.

One important element to data link use with the ATN will be the new-generation digital transceivers that will be required to connect to the ATN. At issue is the type of air-ground and air-air "modem" that will be used. Today's analog radios cannot be used to transmit digital signals at the speeds and accuracy that true ATN will require. A number of development efforts are underway to determine which digital standard should be used for digital VHF communications. One of these is called differential eight-phase shift keying (D8PSK), while another is called Mode-3, or time-division multiple access. The details of these techniques are beyond the scope of this book, but the subject bears watching. The ICAO is expected to make a decision by 1997 on which standard will best meet the ATN requirement.

Applications

Data Link holds the promise of creating a variety of services. Among these possibilities are predeparture clearance (PDC), digital Automatic Terminal Information Service (digital ATIS), and graphical weather. Data link is also considered one of the fundamental technologies necessary to achieve Free Flight. Data link applications will also include ATC services, flight information services, surveillance services, and advanced operational applications such as inflight maintenance fault warnings.

ATC services

Communications between pilots and airspace managers are required to manage the airspace effectively and provide essential air traffic services. Airport operations and airspace management face the limitations of the voice-only communications environment. Data link can augment ATC communications by providing controllers and pilots with a means to transmit the same data digitally. We'll look at some examples of this in a moment.

Data link-based ATC services will be among the first to be implemented in the national airspace system. In one example of controller/pilot data link communications in the en route environment, a controller would be able to create and send an altitude assignment message to the pilot with just a few key strokes and a toggle of the cursor. Controllers and pilots alike can send and receive data link messages. The pilot can respond with WILCO (will comply), STANDBY, or UNABLE. The message is sent via the best available data link back to the air route traffic control center (ARTCC).

The speedy transmission of messages translates into improved aircraft operations and airspace management. A recent FAA study has shown that by reducing ground delays by 62 percent and flight times by almost 20 percent, controller/pilot data link communications in the en route environment may save as much as $340 million annually.

Flight information services

Data link can be used to provide flight information services (FIS) to all participating general aviation and commercial operators. FIS provides general aeronautical information to the pilot to assist in flight planning and management. Currently, a pilot must assimilate a lot of flight information on the ground before flight or in the cockpit via voice and

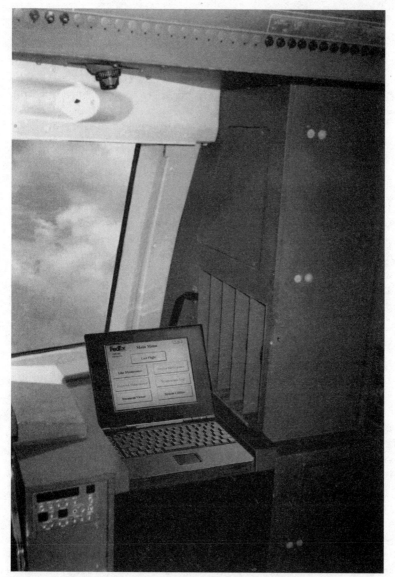

Figure 10.4 Computing Devices' Onboard Maintenance Terminal (OMT) ties a transport-category aircraft's central fault system and the airplane's ACARS. (Computing Devices)

use this information to make decisions. With data link, a pilot could access data visually with the touch of a button—or, ultimately, by voice command to a cockpit communications computer. Data link-enabled FIS could improve cockpit information management, situational awareness, and, ultimately, aircraft safety.

One example of an FIS product is graphical weather displays. With data link, a flight crew could access weather products for their immediate location or any location in the United States with the touch of a button. The display could depict precipitation for the local airport, for example. By use of touch-screen displays, keypads, or trackball pointers, a flight crewmember could zoom in or out to gather more information as the situation required. Data supplied could include NEXRAD weather radar, satellite imagery, terminal forecasts, surface observations, microbursts, and winds aloft.

Surveillance

Aircraft surveillance will be greatly enhanced by data link, by providing controllers and pilots a means of receiving accurate position reports. The two basic methods of data link-enabled surveillance are automatic dependent surveillance (ADS) and automatic dependent surveillance-broadcast (ADS-B). The method of transmission of surveillance data via ADS or ADS-B depends on requirements for surveillance in the particular airspace. Both means provide critical position data that will enhance aircraft operations and management.

ADS

Data link ADS will provide aircraft position information to controllers, airports, and pilots. ADS is just beginning to play a major role in the oceanic environment, where ground-based radar surveillance is impossible and communications are limited to HF radio. Through the use of satcom data link, ADS responds automatically to ground-system requests for aircraft position reports.

In the United States, the FAA is beginning to implement ADS as part of its participation in FANS-1. The FAA's Oceanic Automation Program (OAP) consists of three elements: oceanic ATC, an oceanic traffic management system, and offshore ATC. Oceanic ATC is expected to evolve quickly, and indeed the process has already begun. In October 1995, the FAA began FANS-1 operations at Oakland Center. Initially, 14 aircraft a day were participating in the program, but that number is expected to grow as CPDLC installations continue.

The Oakland ODL system is an enhanced version of the telecommunications processor system that was installed in the Oakland Center earlier in the year. Data link services will be expanded to include the central and northern Pacific regions in late 1996 and to Atlantic Ocean airspace in 1997. A new controller situation display, called OAS, is expected to be installed in 1996, and a complete replacement of the flight data-processing system is underway. Part of the situation display will include a conflict probe capability that will alert controllers when any flight plan or pilot-requested route change will violate any separation criteria. The OAS will eventually be upgraded to the AOAS, which will include features such as a new flight data processor, ADS position reporting, and an enhanced conflict probe.

The Oceanic Traffic Planning System (OTPS) is designed to manage oceanic air traffic using automated information gathering and en route development and analysis. OTPS services will be available at the Oakland, New York City, and Anchorage en route centers and the Air Traffic Control System Command Center (ATCSCC).

Offshore ATC consists of the microprocessor-based En Route Automated Radar Tracking System (EARTS) Phase I and II. Micro EARTS Phase I is the backup for EARTS, which generates the controller's aircraft situation display. According to a GAO report issued in May 1995[1], the existing EARTS equipment has capacity limitations that preclude development and deployment of planned system enhancements. Therefore, Micro EARTS Phase I will be used as the platform for Micro EARTS Phase II, which will provide full EARTS functionality through software enhancements. Micro EARTS, says the GAO report, is an interim replacement for EARTS until a long-term solution is devised. Offshore ATC systems will be installed at the Anchorage, San Juan, Honolulu, and Guam en route centers, as well as at several military installations.

ADS-B

ADS-B is a cooperative system in which each aircraft uses a very accurate navigation system (such as augmented Global Positioning System) and transmits information about its position and, possibly, its intent via an ADS-B channel to the ground and to other aircraft in its vicinity. ADS-B will potentially improve a pilot's situational awareness while enhancing current controller surveillance technology.

In the United States, radar is the primary surveillance tool to provide separation assurance, vectoring, unsafe condition alerts, and traffic advisories. Current surveillance data is obtained via primary and beacon secondary surveillance radar (SSR). While radar has worked well in the past, its limitations (such as resolution, accuracy, and lack of pilot intent) could be inhibiting the evolution of the NAS to include new operational modes such as self-separation and Free Flight. Several ADS-B concepts have been proposed for acquiring surveillance data, but none have so far been demonstrated in a large-scale setting.

Draft requirements for ground surveillance using ADS-B already have been written and should be published soon. Requirements for the air-to-air portion of ADS-B are still in development.

Figure 10.5   Historical artifact: a vacuum-tube-era aircraft radio. (B/CA)

The International Civil Aviation Organization (ICAO) has formed panels to assure worldwide acceptance of ADS-B and agreement on one ADS-B approach to prevent a proliferation of ADS-B standards. Another form of ADS used for oceanic flight, called ADS-Contract (ADS-C), has international standing. All ADS-B concepts need to be demonstrated against real system capacity, interference rejection, robustness, and cost to the user and to the provider.

There are three subsets of ADS-B:

- *Air-to-air ADS-B.* ADS-B also provides air-to-air cockpit display of traffic information (CDTI), a service that provides pilots with surveillance data of all aircraft within the local vicinity. Each aircraft automatically broadcasts its position to all equipped aircraft in the surrounding area; this information is assimilated and visually represented on a display in the cockpit. Since it is not dependent on a ground system, air-to-air ADS-B offers great benefit to many aircraft operators, especially in remote locations. Air-to-air CDTI will also play an important role in the definition of aircraft-based separation, an element of Free Flight. Currently, TCAS provides this function to aircraft that are required to be equipped with TCAS I or TCAS II (discussed in detail in Chapter 2).

- *Air-to-ground ADS-B.* ADS-B provides surveillance data to controllers or aircraft operations facilities on the ground. An aircraft in flight broadcasts its position, altitude, and other pertinent data to ground stations that relay this data to air traffic control or aircraft operations facilities. This data will be used to establish surveillance in remote locations or extend current surveillance capabilities.

- *Ground-to-ground ADS-B.* Ground-to-ground application of ADS-B can also be used to provide surveillance of the airport's surface observations. Aircraft and vehicles broadcast messages containing position, speed, and identification to ground stations around the airport. These messages are relayed to controllers in the tower and personnel in airport management facilities. ADS-B enhanced airport surface surveillance will lead to safer airport surface operations in all weather conditions.

It is expected that ADS-B ground stations will be less expensive to acquire and maintain than radar sites. Since each site will be less costly than today's sites, there is the potential that more sites can be put in service and coverage in remote areas could be improved. In the air-to-air environment, there is the potential for developing an inexpensive method of sending situation data to the cockpit (such as cockpit displays of traffic and intent information and low-cost collision avoidance). Also, since knowledge of the aircraft status is improved with ADS-B, it might be possible to reduce separation standards. Additionally, ADS-B combined with CDTI has the potential to speed up the introduction of Free Flight operations.

ADS-B will also entail the use of local area data link (LADL), which would be a common wireless radio, packet-broadcasting scheme whose protocols are standardized at the data link and physical layers. It would be a kind of wireless Ethernet.

ADS in the cockpit. Honeywell has developed satcom avionics for FANS-1 and has received FAA certification of its FANS-1 software upgrade for the Boeing 747–400 FMS. A slightly different system is currently under development for the Boeing 777

In the cockpits of ADS-equipped Boeing 747–400s, flight crews will continue to communicate with ATC much as they always have. However, instead of using HF radios (with their accompanying drawbacks), the crew can use the aircraft's multifunction control display unit (MCDU) to enter messages, such as requests for altitude changes or course changes to circumnavigate thunderstorms or areas of turbulence. The crew can also simply pick up the cockpit satcom handset and place a phone call, either to ATC or to the airline's dispatch facility.

Conversely, air traffic controllers can contact the flight crew by sending an aural (chime) alert indicating that there is a new ATC message requiring a response. There are also visual attention-getters, such as posting a message in the aircraft's engine indication and crew advisory

Figure 10.6 Cockpits in the next decade will increasingly resemble computer workstations. This development might require meeting new training challenges and using computer-savvy pilots. (Collins)

system (EICAS), which is in the crew's field of view. (This concept isn't new. Selective calling—SELCAL—techniques have been around for years so that crews didn't have to constantly monitor their HF radios waiting for an incoming message.)

Advanced applications. Data link may enhance airspace operations far beyond what can be achieved with voice-only communications. The Future Air Traffic Management/Next Generation (FANG) concept, described shortly, represents the potential for data link to enhance current operations. The Advanced General Aviation Transport Experiment (AGATE) represents one possible future data link-equipped general aviation cockpit. These applications represent a small fraction of data link's potential in the modernization of the national airspace and the advancement of Free Flight.

AGATE. The FAA and NASA, in partnership with the aviation industry, are working together to revolutionize general aviation by integrating new technologies into GA aircraft, ground system, and avionics. The AGATE initiative holds out the potential to increase GA safety, utility, and cost savings with the use of advanced cockpit technologies, including

- Data link
- GPS
- Flat-panel display technology
- Inexpensive microprocessors

Already, however, the marketplace might be showing the way. As I have mentioned elsewhere, pilots of light GA aircraft are often savvy users of personal computer technology, as developers of laptop-compatible GPS receivers and flight-planning software are discovering. Although such systems are not as robust as an ARINC-blessed advanced cockpit suite, they are a far cry from the pencil-and-kneeboard, one-VOR days of recent memory. Indeed, it wouldn't be surprising to find an owner-operator of a Beech Baron or Piper Malibu to be packing an IBM Thinkpad, a portable GPS with a moving-map display, an airborne telephone, and the latest PC flight-planning system in her flightcase.

FANG. FANG is a government/industry research initiative under the direction of the FAA. FANG embraces the concept that we can deliver synergistic system benefits to airspace users by integrating aircraft, airline operation and air traffic management computer systems, and personnel. This kind of integration depends on data link.

FANG will provide aeronautical data link services and requirements for

- Aeronautical data link services
- Requirements for airborne automation systems
- Requirements for ground automation
- Requirements for low-cost aircraft systems to yield partial benefits with minimal equipage

FANG in terminal environment. Data Link can increase capacity in the terminal environment through the automation of descent profile negotiation. Automation systems can compute a fuel-efficient profile to fly an aircraft to a fix at a specific time. However, sometimes conflicts block a preferred path. In these cases, data link can boost efficiency through the automated negotiation of this route.

FANG: Terminal airspace. In the following scenario, FANG is used to negotiate a profile descent in the terminal area.

Step 1: The scheduled required time of arrival (RTA) for a terminal area traffic-metering navigational fix is sent to the aircraft automation system.

Step 2: Upon receipt of this data, the aircraft automation system generates its descent profile trajectory and sends it to ATC through ground automation systems.

Step 3: ATC performs a conflict probe/check and clearance. If there is no conflict, a clearance is sent back to the aircraft. If there is a conflict, a ground automation-generated descent profile is sent to the aircraft. Both these steps are done using data link.

FANG: En route. FANG could enhance many operations in the en route airspace. Frequency tuning, navigation performance, and passing procedures are just a few en route operations that could be enhanced with data link. For example, an aircraft approaching a sector boundary needs to change its radio frequency to that of the next sector for communications purposes. With data link, no voice communications or manual tuning of radios will be required to accomplish this task. Here's the scenario:

Step 1: New frequency up-linked to aircraft from ground automation.

Step 2: Aircraft automation system notifies pilot of new frequency.

Step 3: Pilot accepts radio tuned to new frequency. (This acceptance could be accomplished by pressing an ACKNOWLEDGE button on the data link communications interface.)

Step 4: Aircraft confirms altitude and sends to new sector controller through ground automation.

FANG: En route navigation. The actual navigation performance of an aircraft can be monitored by ATC, and any detected deviations from required navigation performance can be transmitted back to the aircraft all via data link. Again, here's a possible scenario:

Step 1: Aircraft downlink their actual navigation performance (ANP) periodically as they fly along required navigation performance (RNP) routes.

Step 2: The flight management system (FMS) detects navigation-sensing errors and alerts the pilot and ground automation.

Step 3: Ground automation uplinks the corrective action to the FMS.

FANG: En route passing. Passing procedures can be greatly enhanced through data link. The scenario could be:

Step 1: A faster aircraft is flying behind a slower one on the same route.

Step 2: The faster aircraft requests a passing procedure from ATC via data link.

Step 3: The request is granted by ATC, including which way to go and time to initiate the maneuver.

Step 4: The faster aircraft activates the procedure in its flight management system and flies the procedure.

FANG: Oceanic. The oceanic air traffic management system depends on a series of designated tracks for ocean crossings. A problem with this system is the difficulty caused by aircraft entering a scheduled track too early or too late. FANG could solve this operational problem this way:

Step 1: Aircraft A experiences a 30-minute ground delay and loses the reserved slot at the track entry gate.

Step 2: The airline operational control center can request the next available slot from air traffic control via data link.

Step 3: This new track entry time can be sent to the aircraft automation system via data link.

TWIP. The Terminal Weather Information for Pilots (TWIP) represents the first operational-evaluation real-time data link weather products to cockpit. The pilot receives character graphics and text weather products provided by either Terminal Doppler Weather Radar (TDWR) or Integrated Terminal Weather System (ITWS). The TWIP message includes microburst, gust fronts, heavy precipitation, windshear, and storm cell activity within 15 nautical miles of the specified airport. ITWS is a fully integrated weather system to be installed across the United States. Acquiring data from FAA radar and NWS aviation weather products, ITWS will integrate this data for immediate use by controllers and pilots. TDWR is currently being installed across the United States to enhance the safety of air travel through timely detection and reporting of hazardous windshear, microbursts, and gust fronts, as well as wind shifts and precipitation.

During the summer of 1994, USAir, American, Delta, Northwest, UPS, and Federal Express made more than 1000 weather requests per day near the Memphis and Orlando airports. The TWIP demonstration was extended to four additional airports in the summer of 1995 and gained two new participants: Air Canada and AlliedSignal. The FAA plans to make TWIP products available at all TDWR sites within the United States.

GWS.   Graphical Weather Service (GWS) provides real-time graphical weather data to the cockpit. GWS is a nationwide precipitation map derived from a number of ground-based weather radar providers. GWS is currently being evaluated in Frederick, Maryland, using the Dulles Mode S. This evaluation, launched by AOPA Air Safety Foundation and a number of GA user groups, will evaluate the operational suitability and performance of GWS.

TIS.   Traffic Information Service (TIS) is a traffic-alerting system that provides an automatic display of nearby traffic to the pilot in the cockpit. TIS is hosted in the Mode S sensor and uses track reports provided by Mode S surveillance to generate traffic information. Mode S responds to pilot requests with the locations of all of the aircraft within 5 nautical miles. In parallel with GWS, TIS is being evaluated by a number of GA pilots who will provide an operational assessment of the data link products and system.

Ground-to-ground ADS-B.   This operational suitability assessment of data link ADS-B for airport surface surveillance was conducted at Boston's Logan International Airport in early 1994. Four ground stations were used to radiate differential GPS corrections to three equipped vehicles: a Cessna 421, a Cessna 172, and a surface vehicle. The demonstration was a success, providing excellent surface surveillance coverage.

North Atlantic Unified Trials (NUT).   The FAA, in partnership with the United Kingdom, France, Norway, Iceland, Canada, Ireland, and other countries, will sponsor the 1996 North Atlantic Unified Trials, which will demonstrate Oceanic Data Link using the Aeronautical Telecommunications Network (ATN). These trials represent a significant first step toward the implementation of worldwide data link communications.

Air-ground ADS-B: GOMEX.   The use of data link in the Gulf of Mexico (GOMEX) for offshore air-ground surveillance was successfully demonstrated in December 1995. Helicopters servicing oil rigs proved that data link ADS-B can provide surveillance coverage in remote high-traffic areas such as the Gulf of Mexico, where as many as 600 helicopters fly up to 5000 flights per day.

Italian ADS. Alenia of Italy and ARINC will provide an advanced air traffic management system to Aerothai, which operates the Thailand air traffic control system. The contract is for an automated system based on automatic dependent surveillance, with radar surveillance in terminal areas. It has provisions for future controller/pilot data link communications as well as the aeronautical telecommunications network. Alenia will supply advanced ATC automation equipment and integrate it with existing systems in Thailand. ARINC will supply the central processing system and 11 ground stations and the gateway to interconnect the VHF data network with the Alenia equipment. The project is to be completed later in 1996.[2]

Data link, Swedish style. The European Union and civil aviation authorities of several countries are cooperating in an evaluation of the Swedish self-organizing time division multiple access (STDMA) data link for air traffic control. First to use the system in revenue service was a Scandinavian Airlines System Fokker F-28 equipped with a cockpit display of traffic information and the GNSS transponder with an embedded STDMA link. Up to 20 additional commercial aircraft are scheduled to be included in the evaluation. It will consider the effectiveness of the STDMA link for communication navigation surveillance/air traffic management, including precision approaches, air and ground surveillance, the automatic dependent surveillance broadcast mode for collision avoidance, and differential GNSS.[3]

EATCHIP

The transport ministers of 32 countries aligned with the European Civil Aviation Conference (ECAC) have agreed to develop a far-reaching ATC strategy known as EATCHIP (European ATC Harmonization and Integration Program). This initiative calls for wholesale changes in the management of Europe's complex—and often inefficient—collection of airspace fiefdoms, and data link is seen as a major component.

Inflight passenger information systems

While not formally a data link concept, the passenger information and entertainment systems business is seen by many industry experts to be one of the most potentially explosive areas of growth. And some products being developed for this market will rely on a hardware/software architecture that will support large quantities of data being up- and downlinked from aircraft. While current telephony technology is able to keep up (barely) with passenger voice, fax, and (very limited) data-forwarding applications, bandwidth will be the barrier to hurdle before this aspect of data link will become another airline cash cow.

Passengers on long transoceanic flights soon tire of watching movies, and laptop computer batteries are still incapable of crossing the Atlantic in anything slower than Concorde, so product developers are searching for ways to use the passenger seatback and entertainment system console as an entrée to the new interactive passenger cabin service market. They are seeking to develop systems that will enable passengers to link either their own laptops or those rented from the carrier—like earphones currently are—with a local area network within the aircraft. Passengers will have access to a host of standard computer applications, such as word processors, arcade and gameware programs, electronic gambling machines, electronic encyclopedias, Bibles, books, and periodicals, as well as an air-to-ground link to terrestrial communications networks. Such an airborne LAN-based intranet might also alleviate the scientifically unfounded concern that laptops and other portable electronic devices (PEDs) might interfere with the aircraft's navigation or communications systems.

Avionics and data link

Data link is a complex telecommunications environment that depends on the integration of systems on the ground and in the air. The key data link airborne elements required of this interoperable network include

- Control and display unit
- ATN router
- Automation software
- Communications suite

As airborne equipage is essential to complete data link implementation, the FAA is working closely with industry to bring affordable data link avionics to the marketplace.

HF data link. Collins and others in the radio communications field are reexamining HF (high frequency) radio as a possible venue for data link communications. HF first appeared in the 1930s for use as a transoceanic radio communications medium. In the past 10 years or so, HF been superseded by satellite communications (satcom), which offers static-free digital communications links between aircraft, oceanic air traffic control facilities, and the airlines. Standard HF radios have been relegated to the "old tech" category because they are considered to be difficult to use and prone to static from a variety of natural phenomena. However, advances in HF avionics and signal-compression techniques are being incorporated into new designs for high-speed HF radios that will transmit bursts of information (such as maintenance data or passenger communications) that can be more readily captured by high-speed HF receivers. HF could eventually regain its position as a major communications link as traffic growth increases the competition for satcom bandwidth.

C/SOIT: Implementing data link

Implementing the data link telecommunications network will be a complex task, not the least of which is funding the various government elements. The effort has created many industry interdependencies, including

- Integrating data link application software into the future ATC automation systems across all domains
- Establishing a mobile telecommunications "Internet"
- Adopting international standards across the national airspace system
- Developing and producing data link-compatible avionics

Successful data link implementation can only result from a lot of coordinated effort. Toward this end, the FAA has created the C/SOIT (Communications/Surveillance Operational Implementation Team) to address all issues of implementation associated with the task of integrating and deploying a national communications/surveillance system. In cooperation with the aviation industry, the multidisciplinary FAA team has developed packages that contain all the elements necessary for data link implementation in the NAS. Each C/SOIT package addresses the following issues relative to data link's implementation in the NAS:

- Service benefits
- Functional requirements
- FAA support tasking requirements
- FAA approval requirements
- FAA fiscal policy
- Coordination focal points
- Safety and interoperability requirements
- Implementation schedule

Schedule. Data link will be introduced incrementally into the national airspace system. Complex integration with the implementation schedules of the FAA automation systems is required. The implementation can be categorized in three stages: Pre-ATN, ATN, and Advanced ATN. In the current Pre-ATN environment, data link will provide predeparture clearance (PDC), Oceanic Data Link, and Digital-ATIS.

Data link's implementation requires the complex integration with existing and future automation systems, both in the air and on the ground. The advent of digital communications also introduces necessary changes in procedures for pilots and air traffic management. Furthermore, the U.S. implementation of data link must be compatible with the worldwide system to ensure seamless digital communications around the globe.

An FAA effort called Key Site '98 is the FAA's plan to initiate seamless ATN data link communication at key air traffic control facilities in 1998 time frame. These sites would include contiguous tower, terminal, en route, and ocean domains and facilities. This initiative will kick off the national implementation of data link services and communications across all domains.

References

1. General Accounting Office. 1995. *Air Traffic Control—Status of FAA's Modernization Program.* GAO/RCED-95-175FS.
2. *Aviation Week & Space Technology.* 1996. "Filter Center." March 18.
3. Ibid.

Conclusion

*In this work, when it shall be found that
much is omitted, let it not be forgotten that
much likewise is performed.*
—DR. SAMUEL JOHNSON
(on completion of his dictionary in 1755)

Aviation Computing: Poised for Growth

Computers and information networks have revolutionized aviation and probably will continue to do so for some time. Although the diverse disciplines in aviation computing were developed initially to meet very different requirements, they are quickly merging into a seamless environment. The points of interconnectivity are multiplying daily, and one only has to stroll around any of the major aviation gatherings, such as the Paris airshow, the NBAA show, or EAA's annual Oshkosh extravaganza, to see what I mean. Remarkable new computer-based products are being introduced every year that bring new capabilities to recreational, business, military, and commercial pilots, passengers, ATC, and aircraft operators. None of this great stuff was available 20 years ago!

I contend that information technology has, and will continue to, change aviation as significantly as the jet engine did in the 1940s. With few exceptions, the external appearance of aircraft haven't changed very dramatically since the first jet airliners took to the skies in the late 1950s (even the B-2 bomber looks like a throwback to Northrop's XB-49 Flying Wing of the 1940s). And yes, airframe materials, airfoils, powerplants, and flight training have improved as dramatically as the safety record over that same period of time. But the microprocessor has enabled aviation to be transformed from a nuts-and-bolts, slide-rule, electromechanical means of transportation to an activity that enables the flight crew to maximize their resources, the passenger to stay productive (or at least entertained), operators to realize their investments, and aviation authorities to provide a level of safety and efficiency that was inconceivable only 30 or 40 years ago. That's progress, and we should feel pretty good about that.

And things will continue to change. Ultimately, by moving data and applications off the desktop—and possibly out of the cockpit as well—and allowing networks to function as computers, aircraft operators, airframe manufacturers, and air traffic managers can be provided with greater operational flexibility, reduced complexity, simpler upgrades, and less expensive administration. It will come about as a form of synchronicity—developments in aviation information technology seem to be heading, coincidentally, toward a common objective of a seamless, interchangeable field where information can be created almost effortlessly and requested for whatever function is required.

Unfortunately, computers can't provide all the answers. It will be a long time before computers can replace poor judgment, inexperience, and ineptitude. But at the same time, the future holds out some compelling possibilities. The following odds and ends provide a glimpse at some of those possibilities, as well as some questions to consider.

Aircraft design

The designer's draftboard has all but given way to the power of desktop computer-aided design; the production shop floor is incomplete until it is connected to a database—sometimes thousands of miles away—of hypertext assembly instructions and notes. Will the art and craft of creating flying machines give way to a digitized, cost-driven enterprise that could just as easily be producing refrigerators or boom boxes? Has the computer taken away the last vestiges of intuitive art of the designer? I doubt it. There are still plenty of innovative designers out there. Just go to an Oshkosh show if you don't agree. The new-tech tools and materials of the basement aircraft industry are making it possible for more people to get involved in aviation than ever.

Avionics

The flight crews of the future are likely to benefit from advanced technology more dramatically than any other aspect of aviation computing. Continuing development of cockpit automation, combined with advances in sensor and display technology, will make it possible for pilots to operate even more safely and efficiently. But until someone figures out a way to prevent coffee spills on the center pedestal, can all that technology be made completely foolproof?

Romancing the node

For many years, once an aircraft left the gate and took to the skies, it was considered an independent entity. Like a ship at sea, an aircraft is a vehicle under the control of the captain—the pilot in command—until it accomplishes its mission and arrives at its destination. But with the development of a computer-based, data-linked information infrastructure, the aircraft is now being viewed by engineers and designers as a node—a mobile data terminal, producing and digesting, minute-by-minute, information about its location in space, its maintenance status, and the informational and entertainment needs of its passengers. What will access to this new wealth of data do for airlines in the long run? Will it enable the carriers to settle down for a long, uninterrupted period of growth and prosperity? Will we see the return of the bygone romance of air travel? Will today's bottom-line functionality give way to a renewed sense of airline hospitality and style?

Aerospace education opportunity

Where will the next generation of aviation engineers and designers come from? Our culture has changed significantly since many of us used to hang out at the local airport, nurturing a passion for aviation. Now kids hang out on the Internet. Maybe that's a good thing in itself—just learning to navigate the Net is a skill they'll need.

But to capture their imaginations, we'll have to do more than put up a nice home page. The airframe and avionics manufacturers would do well to make their Internet sites "kid friendly"—fun and challenging places where sights and sounds we could only dream about as kids are available at the click of a mouse. Just get kids excited about aviation and aerospace. Show them that not only can they fly magnificent flying machines, but they could someday have a rewarding career designing and testing them, too. How about an aviation student scavenger hunt, where kids could earn points for stopping by aerospace home pages? After earning enough points, the student could trade them in for prizes or free Web access time.

Career opportunities abound

Mike McCann, leader for the integrated product team for the FAA's Information Technology Services, which is part of the Office of Information Technology, told me, "I think the number of career opportunities in the computer field in aviation are just astounding—it's just phenomenal. In the Office of Information Technology we've established a position of chief scientist for software engineering. There is a significant need for expertise in the software and hardware areas, especially as it's related to aviation." These sentiments were echoed almost every-where I went in the aviation industry, and it should send a notice to educators everywhere.

Just a Beginning

So this is the end of our tour, folks. Meanwhile, I hope this book has provided you with what I hoped it would—a sense of the scope and direction of the world of aviation computing. If with this book I have only moved you to sense—as I have—with awe and wonder the startling changes aviation computing systems have produced, then I am satisfied. If this book has helped you decide on a new career direction (hopefully one related to aviation computing), I'm happy. If this book has given your present career in aviation a new perspective and sense of purpose, I'm elated.

The next millennium is upon us. It is obvious to me that specialized, aviation-related information processing systems will continue to shape the aviation industry in ways we still haven't imagined and will take us to the next exciting stage in its evolution. See you there!

Aviation Computing System Resources

The following is a list of organizations whose primary business is providing aviation computing systems and services or, in the case of large companies, have a significant presence in the aerospace industry. Each entry provides standard contact information, product names when known, as well as e-mail and Internet universal resource locator (URL) information. Each entry also has one or more keywords indicating the organization's aviation computing systems specialties. The key below provides an explanation of each keyword. The descriptions are deliberately broad. This information was compiled in September 1996.

Keywords

ADFAR = Airworthiness Directives, FARs, and other technical documents in electronic media (e.g., CD-ROM, online service, software).

BUY$ = Software or service relating to buying or selling aircraft or aviation merchandise.

CAD = Computer-aided design, manufacturing, engineering, document-management software or systems.

CARGO = Air cargo software or data management.

CBT = Computer-based training software or system.

DBASE = Aviation-related electronic databases, software or management service.

FUEL = Fuel-management software and databases.

FBO = Software used for running a fixed-base operation.

INTEG = Aviation software-integration services.

LOG = Aviation (pilot/aircraft/maintenance) logbook and currency-tracking software.

PLAN = Aviation management planning or analysis software or service.

MGMT = Flight department or manufacturing operation software.

MIL = Military-related systems or software.

MX = Maintenance management software.

NAV = Navigation databases and software.

NET = Internet or other aviation network.

NEWS = Aerospace news and information.

REZ = Air carrier or corporate flight department reservations.

SIM = Simulation technology and training.

SKED = Scheduling software.

WX = Aviation weather provider or wholesaler.

F91 = FAR Part 91 and general aviation-related products and services.

F121 = FAR Part 121/127/129/135 air carrier-related products and services.

Accu-Weather, Inc.
619 W. College Ave.
State College, PA
16801
(814) 234-9601 xt. 285
Fax #: (814) 231-0621
Product names: Accu-Data
e-mail: heath@accuwx.com
URL: http://www.accuwx.com
Keywords: WX

Ac-U-Kwik Navpak
P.O. Box 1510
Sausalito, CA
94966
(800) 600-6199
Fax #: (800) 600-6132
Keywords: PLAN, DBASE

Aeroflight Tech Data
P.O. Box 854
Gruver, TX
79040
(800) 756-0650
Fax #: (806) 733-5205
Product names: Aeronautical

Library on CD-ROM and BBS;
  Electronic Logbooks
Keywords: ADFAR, DBASE, FAR

Aeroprice Software
215 South St.
Kingston, MA
02364
(800) 307-0425
Product name: Aeroprice
  Software
e-mail: aeroprice1@aol.com
Keywords: BUY$, DBASE

Air Data Research
13438 Bandera Rd., Ste. 106
Helotes, TX
78251
(210) 695-2204
Fax #: (210) 695-2301
Products: Various aviation safety
  and mechanical reliability
  databases.
e-mail: info@airsafety.com
Keywords: DBASE

Aircraft Dealer's Network
Hangar 3, Wiley Post Airport
Bethany, OK
73008
(405) 491-0803
Fax #: (405) 491-0860
Product name: ADN
  Performance Pro!
Keywords: BUY$

Aircraft Technical Publishers
101 S. Hill Dr.
Brisbane, CA
94005
(415) 330-9500
Fax #: (415) 468-1596
Product names: Numerous titles.
e-mail: info@atp.com
Keywords: ADFAR, DBASE, FAR

Alden Electronics, Inc.
40 Washington St.
P.O. Box 500
Westborough, MA
01581-0500
(800) 225-9492
(508)-366-8851
Fax #: (508) 898-2427
Products: Various
e-mail: info@alden.com
URL: http://www.alden.com
Keywords: WX

AlliedSignal GA Avionics
15001 NE 36th St.
Redmond, WA
98073
(800) 634-3330
(206) 869-6450
Fax #: (206) 869-6464
Product names: AFIS (Airborne
  Flight Information System); Global
  Data Center Flight Information
  Services; CBT training systems for
  flight management systems
Keywords: CBT, PLAN, NAV, WX

Amstat
623 River Rd.
Fair Haven, NJ
77040
(908) 530-6400

Fax #: (908) 530-6360
Product name: AADAM
Keywords: BUY$, DBASE, MGMT

AR Group, Inc.
2925 Briarpark Ave., 6th Fl.
Houston, TX
77042
(713) 430-7000
Fax #: (713) 430-7074
Product name: AiRMET III
Keywords: NAV, PLAN, DBASE

Ashlar
1290 Oakmead Pkwy.
Sunnyvale, CA
90210
(408) 746-1800
Fax #: (408) 746-0749
Product name: Vellum 3D
Keywords: CAD

Autodesk, Inc.
111 McInnis Pkwy.
San Raphael, CA
94903
(415) 507-5000
Products: A variety of CAD,
  mechanical design, data
  management, and multimedia and
  visualization products.
Keywords: CAD

Avantext, Inc.
P.O. Box 369 Reservoir Rd.
Honey Brook, PA
19344
(610) 273-7410
Fax #: (610) 273-7505
Product names: The Airworthiness
  Directive CD; The FAR CD; The
  Aviation Data CD
Keywords: ADFAR, DBASE, FAR

Aviation Aerospace Online
1200 G St. NW
Washington, DC
20005
(202) 383-2398
Fax #: 202.383.2442
Products: Numerous
Keywords: DBASE, NEWS

Aviation Analysis
P.O. Box 3570
Carson City, NV
89702-3570
(800) 736-0392
(702) 882-1011
Fax #: (702) 882-3081
Product names: ADMS III; ADMS IV
e-mail: gary@aviation.com
Keywords: F91, DBASE, FBO, LOG,
   MX, PLAN

Aviation Data Systems
1710 Devon Dr.
Carrollton, TX
75007
(800) 241-2371
(214) 661-5566
Product name: ADS-AIMS (Aircraft
   Information Management System)
Keywords: PLAN, SKED

Aviation Information Systems (AIS)
7455 East Peakview Ave.
Englewood, CO
80111
(303) 779-4787
Fax #: (303) 779-8684
Product name: Dispatch 2001 XG
e-mail: ais@netway.net
URL: http://www.netway.net./ais
Keywords: F91, F121, DBASE, FBO,
   LOG, PLAN, MGMT, MX, SKED

Aviation Software and Analysis, Inc.
   (ASA, Inc.)
244 Bolton Ave.
Beaconsfield, PQ
H9W 1Z8
Canada
(514) 694-3549
Fax #: (514) 694-3549
Product name: Tanker Cranker;
   COTS (Computer Optimized
   Tankering System)
Keywords: PLAN, FUEL

AVTAX Aviation Systems, Inc.
P.O. Box 110
Fairfield, CA
94533-0010
(707) 427-3096
Fax #: (707) 427-3096

Product name: AVTAX
Keywords: MGMT

BACK Information Services
24 Richmond Hill Ave.
Stamford, CT
6901
(800) 446-2225
(203) 974-0650
Fax #: (203) 975-0650
Products: Numerous databases and
   aviation-related analytical support.
Keywords: DBASE, MGMT

Baseops
333 Cyprus Run, Ste. 200
Houston, TX
77094
(800) 333-3563
Fax #: (713) 556-2500
Keywords: F91, DBASE, FBO, PLAN,
   NAV, WX

David Bornemann Associates, Inc.
8133 Leesburg Pike, Ste. 500
Vienna, VA
22182
(703) 821-6848
Fax #: (703) 821-3523
Keywords: F121, DBASE, LOG,
   MGMT, SKED

C.A.L.M.
910 Skokie Blvd., Ste. 112
Northbrook, IL
60062
(800) 545-2256
(708) 480-1133
Fax #: (708) 480-7322
Product names: CALM 2.0; Purchase
   Order; Inventory Management; Bar
   Coding; Wide Area
   Communications; Double Check
Keywords: MGMT, MX

CAE Electronics Inc.
P.O. Box 1800
St. Laurent, PQ
H4L 4X4
Canada
(514) 341-6780
Fax #: (514) 341-7699
Keywords: CBT, SIM

CAMP Systems, Inc.
Long Island MacArthur Airport
999 Marconi Ave.
Ronkonkoma, NY
11779
(516) 588-3200
Fax #: (516) 588-3294
Product names: Numerous titles.
URL: http://www.campsys.com
Keywords: F91, DBASE, FBO, LOG,
 PLAN, MGMT, MX, SKED

Cargo Data Management Corp.
P.O. Box 612146
DFW Airport, TX
75261
(817) 488-3004
Fax #: (817) 488-3303
Product name: Money Track II;
 Station Track II; Easy Track II
Keywords: CARGO, DBASE, MGMT

Cimlinc
1222 Hamilton Pkwy.
Itasca, IL
60143
(708) 250-0090
Fax #: (708) 250-8513/8514
Product name: Linkage Process
 Planning; Linkage Shop View
Keywords: CAD, MGMT

Com-net Software Specialists
3080 S. Tech Blvd.
Miamisburg, OH
45342
(203) 855-9900
Fax #: (802) 422-3840
Keywords: DBASE, INTEG, MGMT,
 NET

Computing Devices International
Aviation Information Systems Div.
3101 East 80th St.
Bloomington, MN
55425
(612) 853-4518
Fax #: (612) 458-4519
Products: Various
Keywords: DBASE, INTEG, PLAN,
 MGMT, MIL, MX, NAV

Compuflight
99 Seaview Blvd.
Port Washington, NY
11050
(516) 625-0202
Fax #: (516) 625-0266
Product name: Compass
Keywords: DBASE, FUEL, FBO,
 PLAN, NAV, WX

Computer Sciences Corp.
2100 E. Grand Ave.
El Segundo, CA
90245
(310) 615-0311
Fax #: (310) 640-2648
Keywords: INTEG, MGMT, MIL

Computer Training Systems (CTS)
580 Sylvan Ave.
Englewood Cliffs, NJ
07632
(201) 567-5639
Fax #: (817) 481-5408
Products: Numerous titles
Keywords: CBT

Computing Technologies for Aviation
 (CTA)
2560 Ivy Rd.
Charlottesville, VA
22903
(804) 971-7624
(804) 971-8041
Fax #: (804) 979-3902
Product name: FOS/2
Keywords: F91, DBASE, FBO, LOG,
 PLAN, MGMT, MX, SKED

Computing Technologies of Virginia,
 Inc. (CTV)
P.O. Box 8256
Charlottesville, VA
22906
(804) 295-8256
Products: Scheduling and tracking
 software for corporate and charter
 operators and aircraft maintenance
 departments.
Keywords: F91, DBASE, FBO, LOG,
 PLAN, MGMT, MX, SKED

Conklin & de Decker Associates, Inc.
P.O. Box 1142
Orleans, MA
02653
(508) 255-5975
Fax #: (508) 255-9380
Product names: ACCESS 2.0; Life
  Cycle Cost Analyzer
Keywords: BUY$, F91, DBASE,
  FUEL, MGMT, MX

Dassault Systemes
1935 N. Buena Vista St.
Burbank, CA
91504
(818) 559-3600
Fax #: (818) 559-3339
Product name: CATIA/CADAM
  Solutions
Keywords: CAD

Data Transformation Corp.
108D Greentree Rd.
Turnersville, NJ
08012
(800) 243-3828 (data)
(609) 228-3232
Fax #: (609) 232-9794
Product name: FAA DUATS (Direct
  User Access Terminal System)
Keywords: DBASE, FUEL, FBO,
  PLAN, NAV, WX

DataWorks Corp.
7441 Lincoln Way
Garden Grove, CA
92641
(714) 891-6336
Fax #: (714) 897-7616
Product names: MAN-FACT II;
  DataFlo
Keywords: CAD, MGMT

Delta Technology International
1621 Westgate Rd.
Eau Claire, WI
54703
(800) 515-6900
(715) 832-7799
Fax #: (715) 832-0700
Product name: Destination Direct

e-mail: deltatec@Primenet.com
Keywords: DBASE, FUEL, PLAN,
  NAV, WX

EMI Aerocorp, Inc.
7 North Brentwood Blvd., Ste. C-10
St. Louis, MO
63105
(800) 951-0006
(314) 727-9600
Fax #: (314) 727-1849
Product names: Aerobrief; Aeronav;
  Aerovor; Aeroroute; Pro-Plan
e-mail: 70007.345@compuserve.com
Keywords: DBASE, FUEL, FBO,
  PLAN, NAV, NEWS, WX

Excel Software Corp.
1072 South De Anza Blvd.
San Jose, CA
95129
(408) 446-5512
Fax #: (408) 446-5513
Product name: TAU (The Aviator's
  Utilities)
Keywords: DBASE, FUEL, PLAN,
  NAV, WX

Flight Data, Inc.
34 South River St.
Wilkes-Barre, PA
18702
(800) 451-3282
Fax #: (717) 824-7698
Product names: DUAT/Plus for
  Windows; Flight Planner for
  Windows
e-mail: flight@epix.net
URL: http://www.flightdata.com
Keywords: DBASE, FUEL, FBO,
  PLAN, NAV, WX

Flight Deck Software
P.O. Box 425
Williamsburg, VA
23185
(800) 955-4359
(804) 229-8516
Product name: IFT-PRO
Keywords: CBT

Flight Level Corp.
2832 Lawtherwood Pl.
Dallas, TX
75214
(214) 327-1001
Fax #: (214) 327-1001
Product name: Flight Level Logbook
  Plus
Keywords: LOG

Flight Watch International
P.O. Box 675259
Marietta, GA
30067
(800) 999-9086
(770) 933-5321
Fax #: (770) 618-3035
Product names: Modules include:
  Scheduling/Operations; Charter
  Quote & Billing; Inventory
  Management; Airline Interface;
  Aircraft Maintenance
Keywords: F91, DBASE, FBO, LOG,
  PLAN, MGMT, MX, SKED

FlightSafety, Inc.
Marine Air Terminal
LaGuardia Airport
Flushing, NY
11371
(718) 565-4100
Fax #: (718) 565-4134
Keywords: CBT, SIM

Global Software Solutions, Inc.
P.O. Box 234
Gardiner, NY
12525
(914) 255-4505
Product names: Aviation Office
  for Windows
e-mail: 73220.3421@compuserve.com
Keywords: FBO, SKED, LOG, PLAN,
  MGMT, SKED

Global Weather Dynamics, Inc.
2400 Garden Rd.
Monterey, CA
93940
(408) 649-4500
Fax #: (408) 649-3521

Product names: Global Weather
  Information Satellite System
  (G*WISS) and Mainframe Service
  Bureau
Keywords: DBASE, FUEL, FBO,
  PLAN, NAV, WX

GTE DUATS
15000 Conference Center Dr.
Box 10814
Chantilly, VA
22021-3808
(800) 345-3828 (help)
(800) 767-9989 (data)
Fax #: (703) 818-5539
Product name: Cirrus
Keywords: DBASE, FUEL, FBO,
  PLAN, NAV, WX

IBM United States
1133 Westchester Ave.
White Plains, NY
10604
(800) 426-4968
(914) 641-5000
Fax #: (914) 641-5065
Product name: CATIA/CADAM
  Mechanical Solutions Ver. 4
  Release 1.5
URL: http://www.ibm.com
Keywords: CAD, MGMT

INAIR, Inc.
18377 Edison Ave.
Spirit of St. Louis Airport
Chesterfield, MO
63005
(314) 530-9109
Fax #: (314) 530-1185
Product name: INAIR
e-mail: 73430.3612@compuserve.com
Keywords: MGMT, MX, SKED

Infax Inc.
2485 Lithonia Industrial Blvd.
Lithonia, GA
30058
(770) 482-2755
Fax #: (770) 482-4057
Product name: WinFIDS
Keywords: DBASE, INTEG, MGMT,
  NET

Jeppesen
55 Inverness Dr. East
Englewood, CO
80112-5498
(303) 799-9090
Fax #: (408) 866-5648
Products: Various flight planning
and weather briefing services and
systems for business, GA, and air
carrier operations.
Keywords: F121, DBASE, FUEL,
FBO, PLAN, MGMT, NAV, WX,
SKED

Kavouras, Inc.
11400 Rupp Dr.
Burnsville, MN
55337
(612) 890-0609
Fax #: (612) 882-4500
Products: Various weather briefing
and flight planning systems.
Keywords: F121, DBASE, FUEL,
PLAN, NAV, WX

McGraw-Hill World Aviation
Directory (WAD)
1200 G. St. NW
Washington, DC
20005
(800) 551-2015
(202) 383-2418
Fax #: (202) 383-2439
Product name: World Aviation
Directory & Buyer's Guide CD-
ROM
Keywords: F121, DBASE,
MGMT

Mentorplus Software
22775 Airport Rd.
Aurora, OR
97002-0356
(503) 678-1431
(503) 678-1480
Product name: FliteStar
Corporate
BBS: (503) 678-1880
Keywords: F91, DBASE, FBO, LOG,
PLAN, MGMT, MX, SKED

Miller Aviation
Binghamton Regional Airport
Johnson City, NY
13790
(800) 288-4228
Fax #: (607) 770-0670
Product name: FBOMIS (FBO
Management Information System)
Keywords: F91, DBASE, FBO, LOG,
PLAN, MGMT, MX, SKED

Mitre Corporation
202 Burlington Rd.
Bedford, MA
1730
(617) 271-2000
Fax #: (617) 271-2271
URL: http://www.mitre.org/
Keywords: INTEG, MGMT

Official Airline Guide
2000 Clearwater Dr.
Oak Brook, IL
60521-8806
(800) 525-1138
Fax #: (708) 574-6070
Product name: OAS Flight Disk
Keywords: F121, DBASE, FBO,
MGMT, REZ, SKED

Pan Am Systems, Inc.
6300 34th Ave. S.
Minneapolis, MN
55450
(800) 367-6602
(612) 727-1084
Fax #: (612) 727-3895
Product name: WeatherMation
Keywords: DBASE, FUEL, FBO,
PLAN, NAV, WX

PCW Aerospace Software
Technology Corp.
261 Berkeley Ave.
Bloomfield, NJ
07003
(201) 743-2908
(215) 672-9177
Fax #: (201) 743-2908
Products: Learjet system simulation
software
Keywords: CBT

Polaris Microsystems
P.O. Box 804
Woodbury, NJ
08096
(800) 336-1204
Fax #: (609) 848-1623
Product name: AeroLog III
Keywords: LOG, MGMT

PRG Aviation Systems
6351 Owensmounth Ave. Ste. 202
Woodland Hills, CA
91367
(818) 710-1425
(800) 877-1425
Fax #: (818) 710-9135
Product names: Flight Operations;
  Aircraft Maintenance; Crew
  Management; Crew Scheduling;
  Inventory management; Line
  Service; General Accounting
e-mail: 75730.52@compuserve.com
Keywords: F121, F91, DBASE,
  FUEL, FBO, LOG, PLAN, MGMT,
  MX, NAV, SKED

Professional Flight Management
  (PFM)
555 E. City Ave.
Bala Cynwyd, PA
19004
(610) 668-1655
Fax #: (610) 668-1624
Product name: PFM Windows
Keywords: F91, DBASE, FBO, LOG,
  PLAN, MGMT, MX, SKED

Professional Software Associates
5746 S. Lisbon Way
Aurora, CO
80015
(303) 699-2090
Fax #: (303) 766-3153
Product names: CAMS (Columbia
  Aviation Management System);
  CRS; DSIS
e-mail: bbrooks@fst.com
Keywords: F121, F91, DBASE, FBO,
  MGMT, MX, REZ, SKED

RMS Technology, Inc.
124 Berkley Ave.
P.O. Box 249
Molalla, OR
97038
(800) 533-3211
(503) 829-6166
Fax #: (503) 829-6568
Product name: Flightsoft
e-mail: 72163.3342@compuserve.com
Keywords: F91, DBASE, FBO, LOG,
  PLAN, MGMT, MX, SKED

SABRE Decision Technologies (SDT)
P.O. Box 619616, MD 4357
DFW Airport, TX
75261-9616
(817) 967-1000
Fax #: (817) 967-9763
Product names: Numerous titles.
Keywords: DBASE, FUEL, FBO,
  LOG, PLAN MGMT, MX, NAV,
  REZ, SKED

SBS International of New York, Inc.
444 Park Ave. S
New York, NY
10016
(212) 532-3230
Fax #: (212) 725-0327
Product names: SBS Crew
  Management System; PAIR;
  LINES; Operations Control
Keywords: DBASE, FUEL, FBO,
  LOG, PLAN MGMT, MX, NAV,
  REZ, SKED

SeaGil Software Co.\BART
PDK Airport
3187 Corsair Dr., Ste. 250
Atlanta, GA
30341
(770) 455-3006
Fax #: (770) 455-1986
Product names: BART 4 Windows;
  BART; Pro Maintenance; Charter
  Quotes
Keywords: F91, DBASE, FBO, LOG,
  PLAN, MGMT, MX, SKED

Silicon Wings, Inc.
4603 Varsity Dr. NW, Ste. 212
Calgary, AB
T3A 2V7
Canada
(403) 286-8804
Fax #: (403) 288-2201
Product: Aircraft inventory control
and maintenance records system
(AIM)
e-mail: cyrj@cadvision.com
Keywords: MX

SimCom Training Centers
7500 Municipal Dr.
Orlando, FL
32819
(407) 345-0511
Keywords: CBT, SIM

SimuFlite Training International
P.O. Box 619119
W. Airfield Dr.
DFW Airport, TX
75261
(214) 456-8000
Fax #: (214) 456-8383
Keywords: CBT, SIM

Software Engineering Associates,
Inc.
153 Country Club Dr.
Box 8899
Incline Village, NV
89452
(702) 832-5333
Fax #: (702) 832-5344
Product name: FlightPak
Keywords: F91, DBASE, FBO, LOG,
PLAN, MGMT, MX, SKED

Spectrum Air Services
8888 West Monroe Rd.
Houston, TX
77061
(713) 645-0070
Fax #: (713) 645-2212
Keywords: F91, DBASE, FUEL,
PLAN, MGMT, NAV, WX

Summit Aviation
P.O. Box 759
Golden, CO
80402
(800) 328-6280
(303) 425-5994
Fax #: (303) 425-7138
Products: FAA regulations,
Airworthiness Directives, Advisory
Circulars, and handbooks on CD-
ROM
Keywords: ADFAR, DBASE, BUY$,
MGMT

Surfware, Inc.
421 Park Ave.
San Fernando, CA
91340
(818) 361-5605
Fax #: (818) 361-1919
Product name: Surfcam
Keywords: CAD

System Resources Corporation
(SRC)
128 Wheeler Rd.
Burlington, MA
18030
(617) 270-9228
Fax #: (617) 272-2589
Products: Numerous
Keywords: INTEG, MGMT

Thompson Training and Simulation,
Inc.
P.O. Box 1677
7041 E. 15th St.
Tulsa, OK
74101
(918) 836-4621
Fax #: (918) 831-2301
Keywords: CBT, SIM

TLC Communications, Inc.
1045 Wildwood Blvd. SW
Issaquah, WA
98027-4506
(206) 392-9592
Fax #: (206) 392-9592
Product name: Air Dispatcher 2.3
Keywords: F91, DBASE, FBO, LOG,
PLAN, MGMT, MX, SKED

Universal Weather and Aviation,
Inc.
8787 Tallyho Rd.
Houston, TX
77061
(800) 231-5600
(713) 944-1622
Fax #: (713) 943-4650
Products: Various
e-mail: info@univ-wea.com
URL: http://www.univ-wea.com/
Keywords: F91, F121, DBASE,
 FUEL, LOG, PLAN, MGMT, NAV,
 WX

World Weatherwatch
401 Bentley St.
Unit 4
Markham, ON
L3R 9T2
Canada
(800) 387-9729
(905) 477-4120
Fax #: (905) 477-0824
Product name: QTABS/Opticom

e-mail: worldwx@accessft.north.net
Keywords: F91, F121, DBASE,
 PLAN, NAV, WX

WSI
4 Federal St.
Billerica, MA
01821
(508) 670-5000
Fax #: (508) 670-5100
Product names: PILOTbrief II;
 WEATHER for Windows
URL: http://www.intellicast.com
Keywords: F91, F121, DBASE,
 PLAN, NEWS, WX

Xionix Simulation, Inc.
8400 Esters Blvd.
Irving, TX
75063
(214) 929-9999
Fax #: (214) 929-0444
Product name: Flight Management
 System Trainers
Keywords: CBT

Aerospace Internet Hot Sites

The following is a selection of some of the many sites which can be found on the Internet. This information was accurate as of September 1996.

Aerospace and Aviation Education

Aviation Safety Reporting System (ASRS) home page (NASA)
http://olias.arc.nasa.gov/ASRS/ASRS.html

Embry-Riddle Aeronautical University
http://www.db.erau.edu/
Links to visitor center, admissions information, faculty, administration, alumni pages, research library, etc.

University of North Dakota Aerospace
http://www.aero.und.nodak.edu/
Links to the UND departments of aviation, atmospheric sciences, computer sciences, space studies, and the UND airline pilot training program, ATC program, aerospace physiology, and extension programs.

Airframe and Powerplant Manufacturers

Airbus Industrie
http://www.airbus.com/

Boeing
http://www.boeing.com
Annual report, division and product descriptions, press releases, company history and company facts, employment opportunities.

Learjet
http://www.learjet.com

Press releases, photos, survey results, company contacts, employment opportunities, gift and apparel shopping, aircraft specifications.

Lockheed Martin
http://www.lmco.com
Links to various products and services, including CASS (Consolidated Automated Support System) course catalog.

The New Piper Aircraft, Inc.
http://www.newpiper.com/
Product and dealer information,
 pilot shop, plant tour information.

Raytheon Aircraft
http://www.raytheon.com/rac/
Overviews, press releases,
 shareholder information,
 employment opportunities.

United Technologies Corp.
http://www.utc.com/
Company overview, news, "Planet
 UTC" corporate tour game.

McDonnell Douglas Corp.
http://pat.mdc.com/
Under construction, March 1996.

Airlines

Aeroflot
http://www.aeroflot.org/Aeroflot.html
Links to Aeroflot Seattle, some of
 Aeroflot's aircraft, Russia/US flight
 schedule

Air France
http://www.airfrance.fr/
Links to products, health
 information, and Concorde

American Airlines
http://www.amrcorp.com/
Links to SABRE, SDT, and AMR
 Corp.

Delta Air Lines
http://www.delta-
 air.com/text/index.html
Links to Delta news, flight plan,

gateways, travelogue, and plane
 fun

Singapore Airlines
http://www.singaporeair.com
One of the most comprehensive
 airline sites in cyberspace.

Trans World Airlines
http://www.twa.com/
One of the least attractive sites I've
 seen; annoying "sign-up" forms,
 links to contact TWA, airport
 information, flights, products and
 services.

United Airlines
http://www.ual.com/
Links to flight information, travel
 tips, mileage plus, etc.

Aviation and Aerospace Associations

Air Transport Association of
 America
http://air-transport.org
Publication catalog; industry data
 and stats; links to ATA member
 services and ATA handbook, which
 describes the airline industry for
 the general public.

Aviation Enthusiast Corner
http://www.brooklyn.cuny.edu/rec/air
 /air.html
Air show listings, museum lists,
 aviation history, a/c locators, WX,
 BBS information.

Aviation Safety Connection
http://www.aviation.org.
A nonprofit organization formed to
 address judgment issues from the
 pilot's perspective; searchable data
 base of NTSB briefs.

Experimental Aircraft Association
http://www.eaa.org
Organization information,
 membership information, Oshkosh
 show information, photo contest,
 more coming . . . a classy site.

Flight Safety Foundation (FSF)
http://rhytech.com/~fsf
[Be sure to enter the tilde (~)]

Flight Test Safety Committee
http://xavier.dfrc.nasa.gov/FTSC/

General Aviation
http://adswww.harvard.edu/GA/
ga_info.html

Inmarsat
http://www.inmarsat.org/inmarsat/
contents.htm
Home page of the international
satellite communications
organization. Explains how satcom
works and where it's headed. Good
stuff!

Micro Wings: The International
Association for Aerospace
Simulations
http://www.microwings.com/index.
html
This is the home page for PC-based
simulator enthusiasts. Message

board, downloadable files, product
reviews, BBS lists and more. Cool!

National Business Aircraft
Association (NBAA)
http://www.nbaa.org
This Web site has both public and
private member access.

National Gay Pilots Association
http://ourworld.compuserve.com:80/
homepages/ron_ngpa/
Links to information packet, local
groups, calendar, Planes We Fly,
and more.

The Society of Flight Test Engineers
http://www.hughes-ec.com/org/sfte/
The site includes a search engine
that yields an abstract of SFTE
papers, which may then be ordered
online.

Aviation Links and Servers

AeroCom
http://www.aero.com

AERONET
http://www.demon.co.uk/aeronet
A collection of airline and business
aviation links from the United
Kingdom

Aircraft Shopper Online
http://www.sonic.net/aso/af.html

Airlines of the Web
http://haas.berkeley.edu/~seidel/
airline.html
Pointers to numerous pax, cargo,
medical airlift airline web sites,
airline stock quotes, 800 numbers,
airline stats, and much more. A
must-see site!

AVWEB
http://www.avweb.com/

Business Aviation Links
http://www.bizjet.com/iban/
Based in the United Kingdom.

GTE Aviation home page
http://www.gtefsd.com/aviation/GTE
aviation.html

The Leading Edge Aviation Hotlist
http://www.fred.net/glen/leading.html

The Web Aviator (Toronto)
http://www.toronto.com/flyer/

U.S. Aviation
http://www.aviationweb.com/links.
htm

Aviation Parts/Spares

Defense Industrial Supply Center
(DISC)-U.S. Navy Aviation Supply
Office (ASO)
http://ebb.disc.dla.mil/

Spare Parts for Buyers
http://www.airparts.com/

Aviation Publications

Aviation Week Group
http://awgnet.com
Links to Internet versions of
*Aviation Week & Space Technology,
Business & Commercial Aviation,
ACFlyer* and other AWG Publica-
tions, as well as Safety/OASIS
aviation safety site.

The Avion Online Newspaper
http://avion.db.erau.edu/

Jeppesen
http://www.jeppesen.com/
Links to company profile, product
catalogs, cool stuff.

Aviation Weather

American Weather Concepts
http://www.amerwxcncpt.com
Aviation WX graphics, text, and
DUAT; recreational/travel WX
graphics; Flyte Trax real-time
information on location of
commercial aircraft; EaasySABRE
reservations; services on
subscription basis.

Accu-Weather
http://accuwx.com

Environment Canada
http://www.dow.on.doe.ca/text/
Provides provincial WX in English
or French.

Universal Aviation & Weather
http://www.univ-wea.com/

University of Florida Agrigator
http://www.IFAS.UFL.EDU/www/
agator/htm/wxr.html
A great WX page!

United Kingdom Wx
http://www.avnet.co.uk

U.S. National Weather Service
http://www.nws.noaa.gov/oso/oso1/
oso12/metar.htm
The new METAR/TAF format goes
into effect on July 1, 1996. Are you
ready? Find out here!

NWS Modernization—You Should
Know
http://cominfo.nws.noaa.gov/
modernize/ysk.htm
Want to know what's new with the
NWS? This is the place to find out.
Worth a visit!

Aviation and Aerospace
Employment Resources

The Airline Employment Service
and The Flight Attendant Career
Guide
http://www.aeps.com/aeps/aepshm.
html

Put your qualifications on file (for a
annual fee); if you are an employer,
use the database to find qualified
applicants.

Avionics

FMS Bluecoat Digest
http://olias.arc.nasa.gov/projects/
bluecoat-digest/registration.html
"Created as a means of opening an
ongoing discussion between the
engineers who build flight deck
automation systems and the pilots
who use them."

Honeywell Commercial Aviation
Systems
http://www.cas.honeywell.com/
Customer service link.

Rockwell-Collins
http://www.cca.rockwell.com

Classic, Antique and Historical

The Online DC-3 Aviation Museum
http://www.centercomp.com/dc3
Lots of information on this

wonderful airplane. Great sound
clips, too!

Classified Advertising

A/C Flyer
http://www.awgnet.com
A/C Flyer is an online magazine
that caters to the world of used
business aviation. Besides sales
listings, you'll find dealer/broker
listings, a product and service
directory, industry news and
editors' picks of other worthwhile
Web sites.

Aircraft Shopper Online
http://www.sonic.net/aso/aso.html
Classified advertisements for a
variety of categories, including
aircraft, fuel, avionics, parts, engines
training, pilot supplies, accessories,
real estate, financing, etc.

Aviation Classifieds
http://haven.uniserve.com/~aircraft/
welcome.html

Federal Aviation Administration

Aviation Safety Reporting System
(ASRS)
http://www-afo.arc.nasa.gov/ASRS/
ASRS.html
Aviation Safety Reporting System
overview, program brief, reporting
forms, ASRS database, immunity
policy, publications, links to related
sites.

FAA important telephone and fax
numbers
http://www.faa.gov/apa/phone/
contact.htm
Need answers to questions? Need to
find someone? This is a good place
to start.

The Official FAA home page
http://www.faa.gov/
This site is another good place to
start tracking down information
about FAA programs, facilities, and
information. Hot links here
include: news and information,
products and programs, the FAA's
information gopher, FAA
information on the World Wide
Web, aviation Internet sites, federal
government sites (Gopher), federal
government sites (WWW), FAA
bulletin board systems, FAA
acquisition reform, and many other
links.

FAA Information
http://aviation.jsc.nasa.gov/faa.html
An "unofficial" home page,
specifically designed for general
aviation use.

FAA Federal Aviation
Administration Aviation Pages
Index
http://www.tc.faa.gov/ZDV/pubs.html
Links to various FAA publications

FAA Center for Advanced Aviation
Systems Development (CAASD)
http://www.caasd.org/

FAA Center for Aviation Systems
Reliability (Iowa State Univ.)
http://www.cnde.iastate.edu/faa.html

FAA Mike Monroney Aeronautical
Center
http://www.mmac.jccbi.gov/

FAA Office of Commercial Space
Transportation
http://www.dot.gov/dotinfo/faa/cst/
ocst.html

FAA Office of System Capacity and
Requirements
http://asc-www.hq.faa.gov/

FAA Program Analysis and
Operations Research Lab (ORLAB)
http://www.orlab.faa.gov

FAA Technical Center
http://www.tc.faa.gov/
Links to general information about
current programs, phone/e-mail #'s,
FAA grants and contract
solicitations, aviation education,
administrator's fact book, employee
programs, an FTP server and more.

Military Aviation

U.S. Air Force
http://www.af.mil
Links to news, fact sheets, people,

Milwaukee Airport Traffic Control
Tower
http://www.faa.gov/ats/mkeatct/
mke_home.htm

FAA NORCAL
http://www.aero.com/publications/
norcal/norcal.htm or try:
http://www.armory.com/norcal/
FAA SOCAL
http://www.cco.caltech.edu/~aacit/
faa_seminars.html

Regulation and Certification (AVR)
group
http://faa.gov/avr/avrhome.htm

Human Factors Program
http://www.faa.gov/aar/human-
factors/welcome.htm

Mitre Corporation
http://www.mitre.org/
Nonprofit corporation that develops
systems for the DOD and FAA.

MIT Lincoln Labs
http://www.ll.mit.edu/
Applied science lab for advanced
technology, including ATC and
other aerospace-related subjects

MIT Lincoln Labs Programs
http://www.ll.mit.edu/Links/
programs.html
Current aerospace programs

NASA Langley Research Center
Simulation Systems Research
Branch
http://bigben.larc.nasa.gov/branch/
branch.html
Links to various programs at the
SSB. Lots going on here!

pictures, overview, other sites. A
must stop on a tour of aviation on
the Internet.

BosniaLink
http://www.dtic.dla.mil/bosnia
DOD Internet WWW page to
support our servicemen and
servicewomen in Bosnia and to
publish information about U.S. and
NATO activities

U.S. DOD Defense Technical
Information Center home page
http://www.dtic.dla.mil/
Links to many other sites.

The Official NATO home page
http://www.nato.int/

U.S. Navy
www.navy.mil
Links to FAQ answers, Blue Angels
show schedule, Navy news, public
affairs, recruiting, and Internet
dial-up services.

The Royal Air Force (UK)
http://www.open.gov.uk/raf/rafhome.
htm
[Please note that htm is not input as
html.]

Miscellaneous

AIAA home page
http://www.lainet.com/~rich/
aiaaonline/

Air & Space Magazine
http://airspacemag.com

AOPA-UK
http://indus.easynet.co.uk/aopa/

ARINC, Inc.
http://www.arinc.com

Aviation resources & fire equipment
images
http://ids.net/~slemay/home.html

Computing Devices Intl.
http://www.cdev.com/
Commercial and military data-
processing systems developer.

Duncan Aviation
http://www.infoanalytic.com/
duncan/welcome.html

Hubble Space Telescope's Greatest
Hits 1990–1995
http://www.stsci.edu/pubinfo/BestOf
HST95.html
Mind-blowing pictures from deep
space. A must-visit!!!

Jane's home page
http://www.janes.com/janes.html

NACA Server
http://www.larc.nasa.gov/naca/naca.
html
Listings of NACA reports.

Neil Krey's home page
http://ourworld.compuserve.com/
homepages/NeilKrey/
Papers and presentations which
have been written over the years
on CRM and cockpit automation.
Material on simulation, learning
theory, virtual reality and distance
learning to be added soon.

Sky Warriors
http://www.skywarriors.com
Home page of the air combat
re-creation folks.

US Geological Survey declassified
spy satellites photos
http://edcwww.cr.usgs.gov/dclas

Wonderful World of Flying
http://www.wwof.com

World Aircraft
http://trex.smoky.ccsd.k12.co.us/
~dlevin/air/air.html

Glossary

"Avcomps" Contraction of *"Aviation Computing Systems"* (or *Services*), coined by *Business & Commercial Aviation* magazine in the early 1980s.

3-D animation tools Software to help engineers visualize objects on CAD/CAM systems.

604 Circuit Distribution network for weather and other aeronautical data to airlines, FAA facilities, and third-party weather vendors.

AAS Advanced Automation System.

AC Advisory Circular (FAA).

ACARS Aircraft communications and reporting system; developed by ARINC for air carrier purposes.

ACCC Area Control Computer Complex.

ACFP Advanced Computer Flight Plan (military).

AD Airworthiness Directives; FAA certification directives having the legal power of the FARs.

Ada Avionics programming language developed by the DOD; named after Augusta Ada Byron, daughter of Lord Byron and friend of Charles Babbage, who is credited with developing the first mechanical computing machine.

Adaptive flight displays Avionics displays systems capable of displaying different types of information, depending on the mode selected by the pilot.

ADNS ARINC Data Network Service.

ADS Automatic dependent surveillance; includes a variety of radar- and satcom-based air/ground and air/air traffic separation technologies that are considered key elements of the Future Air Navigation System (FANS).

ADS-B Automatic Dependent Surveillance-Broadcast, a cooperative form of air/air separation that relies on very accurate navigation systems (e.g., GPS) and frequent position broadcasting.

ADTN-2000 Administrative Data Telecommunications Network; the FAA's in-house administrative communications network.

AEEC Aircraft Electronic Engineering Committee.

AERA Automated En Route ATC.

AERO-H   Satcom services provided by Inmarsat for major air carriers and long-range corporate aircraft requiring long-range, high-speed communications.

AERO-I   Satcom services provided by Inmarsat for use by regional airlines and corporate aircraft.

AES   Aircraft earth station. The airborne radios used for satcom.

AFGWC   Air Force Global Weather Central.

AFIS   Aviation Flight Information System; an ACARS-like air/air and air/ground communications network developed by AlliedSignal; used primarily by operators of business aircraft.

AFIS   Airborne Flight Information System.

AFSFC   Air Force Space Forecast Center.

AFTN   Aeronautical Fixed Telecommunications Network.

AGATE   Advanced General Aviation Transport Experiment.

AGFS   Aviation Gridded Forecast System (FAA).

AIMS   Aircraft Information Management System (Boeing 777).

Air tasking orders   Instructions for carrying out airborne military operations.

ALCM   Airport Lighting Control Monitor.

ALPA   Air Line Pilots Association.

ALS   Autonomous landing system.

AME   Aviation Medical Examiner.

AMLCD   Active-matrix liquid crystal display.

AMOSS   Airline Maintenance and Operations Support System (Honeywell).

ANMET   Antilles Meteorological Network.

ANSI   American National Standards Institute.

AOA   Angle of attack (aerodynamics).

AOAS   Automated Oceanic ATC System.

AOPA   Aircraft Owners and Pilots Association.

Apollo/Galileo/Sierra   Air carrier computerized reservations system.

ARINC   Aeronautical Radio, Inc.

ARTCC   Air route traffic control center.

ASCB   Avionics Standard Communications Bus (Honeywell).

ASD   Aircraft Situation Display.

ASDS   Aviation Support Data System (military).

ASIP   AFSFC Software Improvements Program (military).

ASIS   Aviation Standards Information System.

ASOS Automated Surface Observing Systems.

ASRS Aviation Safety Reporting System.

ATA Air Transport Association.

ATCRBS Air Traffic Control Radio Beacon System.

ATCSCC Air Traffic Control System Command Center; FAA's central traffic management hub, based in Herndon, VA, also called Flow Control.

ATIS Airport Terminal Information System.

ATMS Advanced Traffic Management System.

ATN Aeronautical Telecommunications Network

ATPCO Airline Tariff Publishing Company.

AWAC Airborne Warning and Control aircraft.

AWDS Aviation Weather Distribution System.

AWIPS Advanced Weather Interactive Processing System.

AWN Automated Weather Network (Military).

AWOS Automated Weather Observing System.

AWPG Aviation Weather Products Generator.

AWS Air Weather Service (military).

BBS Electronic bulletin board system.

BDA Bomb damage assessment.

BWS Base Weather Station (military).

C/SOIT Communications/Surveillance Operational Implementation Team (FAA).

C3I Command, control, communications, and intelligence.

C4I Command, control, communications, computers and intelligence.

CAA Civil Aeronautics Authority (United States).

CAA Civil aviation authority or agency.

CAB Civil Aeronautics Board.

CAD/CAM Computer-aided design/Computer-aided manufacturing.

CAE Computer-aided engineering.

CAIS Comprehensive Airmen Information System.

CATIA A type of CAD/CAM software developed by Dassault Aeronautique and later licensed to IBM for marketing purposes.

CBT Computer-based training.

CCIP Constantly computing impact point.

CD-ROM Compact disk, read-only memory.

CDU   Control display unit.

Center   Another term for ARTCC.

CFD   Computational fluid dynamics; computerized visualization of airflow patterns near an aircraft in flight; used for creating aerodynamic structures.

CFEP   Communications Front End Processor (military).

CMET   Central American Meteorological Network.

CNS   Communication, navigation, surveillance (ATC).

CNS/ATM   Communication, navigation, surveillance/air traffic management.

Codec   Encode/decode (satcom).

Conflict probe   A rules-based or computerized method for determining whether two or more aircraft will remain separated by adequate margins, and an aid for maximizing the efficiency of airspace; a key element of Free Flight.

CORN   Computer Resource Nucleus.

COTS   Commercial off-the-shelf systems.

CPDLC   Controller-pilot data-link communications (ATN).

CRM   Crew resource management.

CRS   Computerized reservations system (air carrier).

CRT   Cathode ray tube.

CSD   Computational stress design; used to determine stress loads on structures.

CSDB   Collins Commercial Standard Digital Bus (Rockwell-Collins).

CWIN   Cockpit Weather Information Network.

CWS   Combat Weather System.

D8PSK   Differential eight-phase shift keying (ATN).

DARC   Direct Access Radar Channel.

DARP   Dynamic Airborne Route Planning (data link).

DCC   Display Channel Complex.

DCCR   Display Channel Complex Rehost.

DCLS   Digital control loading system (simulators).

DEA   Drug Enforcement Agency.

DGPS   Differential Global Positioning System.

DME   Distance measuring equipment.

DMS   Digital motion system (simulation).

DOD   U.S. Department of Defense.

DSP   Digital signal processing.

DSR Display System Replacement.

DTM Digital transfer modules.

DUAT or DUATS Direct User Access Terminal System (FAA).

E6B "whiz wheel" A hand-held, circular mechanical slide rule used for calculating a variety of navigational and flight-planning problems.

EAA Experimental Aircraft Association.

EADI Electronic altitude direction indicator.

EARTS En Route Automatic Radar Tracking System (ATC).

EATCHIP European ATC Harmonization and Integration Program.

EFIS Electronic flight information system.

EHSI Electronic horizontal situation indicator.

EICAS Engine indication and crew advisory system.

ELS Electronic library system.

EM Electromechanical avionics.

ERP Enterprise resource planning.

ETMS Enhanced Traffic Management System.

ETOPS Extended twin-engine overwater operations.

EVS Enhanced vision system.

FADE/SIM FAA-Airlines Data Exchange/Simulation.

FADEC Full Authority Digital Engine Control.

FANG Future Air Traffic Management/Next Generation.

FAR Federal Aviation Regulation.

FAR/AIM Federal Aviation Regulation/Aeronautical Information Manual; when seen together, this often refers to a CD-ROM or database containing these items together. Typically included is a search engine software.

FBO Fixed-base operator.

FIDS Flight information display system.

FIR Flight Information Region; airspace in international areas, normally administered by one or more national ATC organizations.

FLIR Forward-looking infrared radiometer.

FMC Flight management computer.

FMS Flight management system.

FSDO Flight Standards District Office.

FSL Forecast Science Laboratory (NOAA).

FSS Flight Service Station.

FTD   Flight training device.

FTP   File transfer protocol.

FWOS   Fixed Weather Observing System (military).

GA   General aviation.

GAO   General Accounting Office.

GDM   Ground Delay Manager.

GES   Ground earth station. The terrestrial stations that send and receive satcom transmissions.

GNSS   Global navigation satellite system.

GOES   Geostationary Operational Environmental Satellite.

GOMEX   An experimental data link system tested in the Gulf of Mexico.

GPWS   Ground Proximity Warning System.

GTS   Global (Meteorological) Telecommunications System.

GWS   Graphical Weather Service.

HARS   High Altitude Route System.

HARV   High Angle-of-Attack Research Vehicle.

HF   High frequency; the portion of the radio spectrum from 2 to 30 megahertz. Used by transoceanic aircraft for communications purposes. A pulsed, digital form is currently in development as an alternative to satcom.

HSCT   High-speed civil transport.

HTML   HyperText Markup Language.

HUD   Head-up displays; used on a growing number of civil and military aircraft, HUDs provide images of symbols representing flight instrument information such as speed, heading altitude, trend vectors, and an artificial horizon. The images are superimposed on a holographically sensitized visor that the pilot looks through to see through the windscreen. The advantage is that the pilot can remain oriented with the outside view while being aware of critical flight information. This is especially helpful when conducting instrument approaches.

HUMS   Health usage monitoring system (helicopters).

IC   Integrated circuit.

ICAO   International Civil Aviation Organization.

ICO   Intermediate Circular Orbit (satellite).

IFCS   Integrated Flight Control System.

IFE   In-flight entertainment system.

ILS   Instrument landing system. The primary ground-based precision landing system used throughout the world; provides lateral and vertical guidance.

Inmarsat   International Mobile Satellite. A London-based organization comprised of representatives of nations that operate mobile (land, sea, air, portable) satcom systems for users in their countries and other countries.

ISSS   Initial Sector Suite System (ATC).

ITWS   Integrated Terminal Weather System (FAA).

JAA   Joint Airworthiness Authority.

JMEM   Joint Munitions Effectiveness Manual.

LCD   Liquid crystal display.

LEO   Low earth orbit (satellite).

LOFT   Line-oriented flight training.

LSO   Landing Signal Officer (military).

Massport   Massachusetts Port Authority.

MCDU   Multifunction control display unit.

MDAU   Maintenance data acquisition unit.

Medevac   Medical evacuation.

MEL   Minimum equipment list.

METAR   Meteorological Reports, Airdrome. Hourly reports on conditions at airports and other landing areas.

MFD   Multifunction display.

Mil-Spec   Military specification.

MIL-STD   Military standard.

MIS   Management Information System (or specialist).

MNPS   Minimum navigation performance standards. Standards established by ICAO signatories to ensure that aircraft operating in remote or transoceanic areas are using navigational systems that are suitable for those operations.

MOA   Military Operating Area.

Modem   Modulate/demodulate.

MPS   Mission Planning System (military).

MUFIDS   Multiple-User Flight Information Display System.

NADIN   National Aviation Data Interchange Network.

NAS   National Airspace System (FAA).

NAS   Naval Air Station.

NASA   National Aeronautics and Space Administration.

NAT   North Atlantic Tracks. A changing set of navigational fixes and altitude assignments used by the ATC for aircraft crossing the north Atlantic Ocean.

Routes are assigned on an east-west system to take best advantage of prevailing upper-air winds on a given day.

NATCA   National Air Traffic Controllers Association.

NATO   North Atlantic Treaty Organization.

NAVOCEANCOM   U.S. Naval Oceanography Command.

NAWPP   National Aviation Weather Program Plan.

NBAA   National Business Aircraft Association.

NCSA   National Center for Supercomputing Applications.

NEDN   U.S. Naval Environmental Data Network.

NEXRAD   Next-generation radar.

NMC   NOAA National Meteorological Center.

NOAA   National Oceanographic and Atmospheric Administration.

NOAAPORT   NOAA Data Transmission System. Destined to become part of the AWIPS program.

NOMSS   U.S. Naval Oceanographic and Meteorological Support System.

NOS   National Ocean Survey.

NRP   National Route Program.

NSC   National Simulation Capability.

NSSFC   National Severe Storms Forecast Center.

NTSB   National Transportation Safety Board.

NTSC   Volpe National Transportation Systems Center, Cambridge, Massachusetts.

NUT   North Atlantic Unified Trials (ATN).

NWS   National Weather Service.

OAG   Official Airline Guide.

OAP   Oceanic Automation Program.

OAS   Oceanic ATC System.

ODL   Oceanic Data Link.

OEM   Original equipment manufacturer.

OOOI   Out, off, on, in (air carrier). For dispatch and management purposes, flight crews or electronic equipment installed in the aircraft record the time an aircraft leaves the gate (out) (parking brake release time), leaves the ground (off), returns to the ground (on), and arrives at the destination gate (in) (set parking brakes).

OOPS   Object-oriented programming software.

OPS   Onboard Performance System.

PAMA   Professional Aviation Maintenance Association.

PAMRI   Peripheral Adapter Module Replacement Item (ATC).

PC-FTD   Personal computer flight training device.

PCS   Power conditioning system (ATC).

PDC   Predeparture clearance.

PDM   Product data management.

PED   Portable electronic devices (such as laptop PCs, radios, and CD/tape recorders).

PFD   Primary flight display.

POI   Principal Operations Inspector (FAA).

PRAT   Prediction/Resolution Advisory Tool (Free Flight).

PRM   Precision Runway Monitor (ATC).

Protocol   A method for two computers to talk to each other.

RIDS   Ramp information display system (air carrier).

RISC   Reduced instruction set computers.

RPM   Revenue passenger miles (air carrier).

RVSM   Reduced Vertical Separation Minimums (ATC).

SABRE   Semi-Automated Business Research Environment (air carrier CRS).

SELCAL   Selective Calling; a VHF/HF system to enable a ground-based operator (Company or ATC) to contact an individual flight; relieves flight crews from having to monitor radios for calls on long, oceanic flights.

SID/STAR   Standard instrument departure/standard terminal arrival route.

SITA   Societe Internationale de Telecommunications.

STDMA   Self-organizing time division multiple access (ATN).

SVS   Synthetic vision system (avionics).

TAF   Terminal Airport Forecast. Weather forecasts for certain airports.

TCAS   Traffic Alert Collision Avoidance System.

TCCC   Tower Control Computer Complex.

TCP/IP   Transmission Control Protocol/Internet Protocol.

TDMA   Mode-3 time division multiple access (ATN).

TIS   Traffic Information Service.

TOAST   Terminal Operations and Schedule Tracking (Comair).

TOT   Time on target (military).

TPFDF   Transaction Processing Facility Data Format.

TRACON   Terminal radar approach control (ATC).

TSRV   Transport Systems Research Vehicle (NASA).

TWIP   Terminal Weather Information for Pilots.

UAV   Unmanned aerial vehicle.

URL   Universal resource locator (Internet).

VHF   Very high frequency.

VNAV   Vertical navigation (avionics).

VOR   Very high frequency omnirange radio (navaid).

VSCS   Voice Switching Control System (ATC).

WAAS   Wide Area Augmentation System (ATC).

WAC   World Aeronautical Chart.

WAFC   World Area Forecast Center.

WAFS   World Area Forecast System.

WAN   Wide area network.

WETMS   Weather teams (military).

WFO   Weather Forecast Office.

WMO   World Meteorological Organization.

WSU   Weather Support Units (military).

WWW   World Wide Web.

X.25 and X.75   Messaging protocols used for packet-switching-based communications networks such as the Internet.

Index

ABOUT THE AUTHOR

Mal Gormley began flying in 1977, while working in the
Readers Digest Association's computer center. Subsequently,
he earned his Airline Transport Pilot rating while working
as a flight instructor, charter pilot, and FAA-designated
Pilot Examiner; he later flew for a commuter airline. In
1988, Mal joined the staff at *Business & Commercial Aviation* magazine, where he specialized in corporate aviation
topics, including aviation computing and Satcom. In 1994,
Mal moved his family to Maine and began to work as a freelance contributor to *B/CA, Aviation Week & Space Technology,* and other publications. He helped to launch the
Aviation Week Group Information Center on CompuServe,
and developed the *Safety/OASIS* portion of Aviation Week
Group's Internet Home Page.